Logistic Core Operations with SAP

Jens Kappauf · Bernd Lauterbach
Matthias Koch

Logistic Core Operations with SAP

Procurement, Production and Distribution Logistics

 Springer

Jens Kappauf
Heidelberg
Germany
info@books.kappauf.eu

Bernd Lauterbach
Heidelberg
Germany
b.lauterbach.logistics@googlemail.com

Matthias Koch
Karlsruhe
Germany
m.koch.logistics@googlemail.com

ISBN 978-3-642-44658-0   ISBN 978-3-642-18204-4 (eBook)
DOI 10.1007/978-3-642-18204-4
Springer Heidelberg Dordrecht London New York

Printed on acid-free paper

Springer is part of Springer Science+Business Media (www.springer.com)

Contents

Chapter 1
Introduction

The most imperative challenge facing a number of managers in recent years has been to regain lost market share and secure new competitive advantages. The impetus for this trend continues to be the ubiquitous tendency toward globalization and the ensuing intensification of international competition. *Customer orientation*, *lean management* and *re-engineering* are the buzzwords that characterize these efforts.

In fact, several companies need to reorganize their value-added processes, whereby special attention should be paid to the interfaces between the sales and procurement markets, which are increasing in importance. Within this context, there is hardly a corporate function that has grown in significance in recent years as much as *logistics*.

Treated until just a few years ago as an operational aid and an object of isolated rationalization efforts, logistics—especially in the age of *supply chain management*—now is considered an essential element of strategic corporate leadership. More and more, logistics is being functionally mapped in standard business software. Accordingly, there is great demand for logistical expertise in connection with the know-how surrounding the implementation of logistics in complex information systems.

> **Definition of Supply Chain Management**
> *Supply chain management* (SCM) is the observation and administration of logistical processes along the entire value creation chain, which includes suppliers, customers and end consumers.

We have divided our presentation of logistics operations with SAP into two volumes. The purpose of them is to provide you with an introduction to the world of logistics with SAP software and assist you in understanding the terminology, concepts and technological components as well as their integration. Because the described processes are complex and include a number of functional details, we

J. Kappauf et al., *Logistic Core Operations with SAP*,
DOI 10.1007/978-3-642-18204-4_1, © Springer-Verlag Berlin Heidelberg 2011

have attempted to make the examples as representative as possible in terms of the presentation and functional explanation of the SAP system components, SAP ERP (Enterprise Resource Planning) and SAP SCM (Supply Chain Management). This means that we will discuss all components of SAP Business Suite and core functions within the context of logistics. A few components, especially technical ones, and functional areas (e.g. disposal, maintenance, compliance and service management) will not be covered.

We have taken care to explain business-related questions and their SAP-specific solutions as well as technical terms, and illustrate their relationships. The books are designed to provide an easily understandable yet well-substantiated look at the respective process chains, and be a useful source of information for everyone— from IT experts with only basic knowledge of the business-related issues, to employees in logistics departments who are not yet familiar with SAP terms and applications.

1.1   Whom Do These Books Address?

Logistic Core Operations with SAP cannot answer every query, but we hope to give you the tools with which to ask the right questions and understand the essential issues involved. The contents of this book are thus aimed at the following target groups:

The books are dedicated to everyone looking for a lucid, informed introduction to logistics with SAP. Thus, each chapter describes in detail a specific logistics field and provides an overview of the functionality and applications of the respective components in practical business use. In this regard, we address SAP beginners and employees in departments where SAP is to be implemented, as well as students wishing to obtain an impression of the logistic core processes and their mapping in SAP software.

We also speak to ambitious users of SAP Business Suite who, in addition to relevant logistical processes, want a look at process integration and the functions up- or downstream, as well as their mapping in SAP Business Suite.

Last but not least, we turn to management staff and IT decision-makers who are considering the implementation of SAP Business Suite or its individual components and wish to obtain an overview of logistical processes with SAP systems.

1.2   Operational Significance of Logistics

The operational significance of logistics for many companies still lies in its rationalization potential. In general, a reduction of logistics costs should improve corporate success by achieving a competitive advantage. Surveys of businesses have demonstrated that, for the coming years, companies are still counting on a considerable cost-reduction potential of 5–10% of total costs (see 3PL Study 2009).

This statement does not contradict the fact that the share of logistics costs of many companies was more likely to increase in the past because, for instance, it is directly related to which operational processes are included in the logistics process.

Thus, the scope of logistics in recent years has continually expanded, for example to include production planning and control (PPS systems) or quality control. In addition, significant investments are being made in IT technology, in areas such as supply chain management planning. In the near future, this will lead to a decrease in administrative logistics costs (e.g. through shipment tracking, transport organization or Internet-based ordering).

Further savings are expected in commercial and industrial firms by subcontracting logistics services (logistics outsourcing). In particular, operative logistics tasks, such as transport, storage, commissioning and packaging, have already been outsourced to a high degree to external logistics service providers. However, since a lack of quality in logistics services is generally not blamed on the service providers involved, but rather on the supplier, outsourcing logistics functions can be problematic.

When the quality of competitor products continues to become more comparable and there is hardly room to lower prices any further, competition takes place on the level of service performance. Within these services, logistics ranks highly: Delivery dependability, rapid returns processing and a high degree of customer service quality are characteristics with which a company can set itself apart from its competitors.

Several logistics processes either include interfaces with customers or have effects on the customer. That is why logistics processes must be oriented toward customer needs and performed in a service-friendly manner. In an era when logistics demands are becoming ever more exacting and, by the time the consumer is reached, ever more customized, companies that master these processes to the advantage of their customers will experience a competitive edge that, at least in the short term, cannot be bridged by the competition. Firms boasting excellent logistics management can hardly be replaced by other suppliers. Thus, there are cases in which logistics in commercial and industrial businesses is one of the core competences for which outsourcing should *not* be considered. This is not to say that the fulfillment of basic logistics functions (e.g. transport or storage) cannot be outsourced, as there are plenty of providers on the market that are capable of taking over the logistics tasks of a previous supplier on a short-term basis without detriment to quality (make-or-buy decision).

There are a growing number of ever-changing definitions and classification options offered in print as well as on the Internet for the term *logistics*. Of these, we would like to use the functional, flow-oriented definition of the American logistics society "Council of Supply Chain Management Professionals" as the basis for this book and its journey into the logistics options of the SAP Company:

> Logistic management is that part of supply chain management that plans, implements, and controls the efficient, effective forward and reverse flow and storage of goods, services, and related information—between the point of origin and the point of consumption in order to meet customers' requirements. (Source: Council of Supply Chain Management Professionals)

According to this definition, logistics serves to move goods within the entire value chain and requires coordination and integration between companies. It focuses primarily on real goods, tangible assets and services that provide benefit to the customer, and it integrates them into the core logistics functions of transport, transfer and storage.

Logistics therefore comprises the planning, control and execution of goods and information flow—between a company and its suppliers, within a company, and between a company and its customers.

Materials management, on the other hand, includes all activities involved in supplying a company and its production processes with all necessary materials at optimal cost. Logistics takes into account spatial and temporal gaps involved in supply processes, not only with regard to the materials, but also to the information to be exchanged between business partners. That is why we consider materials management to be not only a part of logistics, but its center, whereby the functions of logistics are more comprehensive than those of materials management.

A further possibility of logistics classification is the differentiation of logistics phenomena according to functional aspects. As a cross-sectional function, logistics maintains interfaces with the primary functional business areas of acquisition, production and sales.

We thus traditionally differentiate between the following logistics areas in the order in which goods flow through a company, from the acquisition to the sales market:

- Procurement logistics
- Production logistics
- Distribution logistics

Current logistics definitions augment these traditional core areas to include further aspects. These include disposal logistics and operational maintenance, or service management. Spare parts logistics ensures the materials-management-related supply and availability of spare parts.

With this project, we have attempted to provide as comprehensive a portrayal of logistics processes and issues as possible, which include not only theoretical principles but also problems of practical use, as well as their implementation in SAP Business Suite. Therefore, we have expanded the traditional cross-sectional business functions to include the following logistics areas, to which chapters in Volume II are dedicated:

- Transport logistics
- Warehouse logistics and inventory management

Due to our goal of providing basic knowledge of logistics core processes and their mapping in SAP Business Suite within the framework of conceptual possibilities, disposal logistics and service management and maintenance (as well as compliance) are not within the scope of this book. For more information on these topics, refer to the bibliography. There you will find information concerning all of the books or sources we have quoted or referenced.

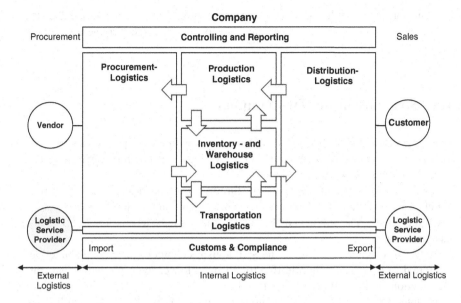

Fig. 1.1 Functional areas of logistics

Figure 1.1 shows the classic and expanded functional areas of logistics that will be discussed in more detail in *Logistic Core Operations with SAP*.

On the side of the *procurement markets,* it is the task of *procurement logistics* to acquire the articles as well as raw materials and supplies necessary for the operational processes of manufacturing and distribution. Procurement is carried out with reference to a certain procurement and stock situation, especially based on materials management planning as part of *production logistics*. The result of such planning may be a purchase requisition. The purchase requisition is cleared for procurement, converted into an order and transferred to the determined source for internal or external procurement. The conclusion of a procurement transaction may involve receiving and paying a supplier invoice, in addition to goods receipt into the warehouse. The Goods Receipt Department not only records stock but also its valuation for the Accounting Department. Transfer of the materials to stock, their quality inspection and inventory management are part of *warehouse logistics* and *inventory management.*

Distribution logistics primarily concerns sales processes that generally begin with a customer ordering materials and indicating a desired delivery date. Using this information, a sales order is generated. Depending on the delivery date, shipping activities are initiated such that the materials reach the customer in a timely manner. *Warehouse logistics* takes over the task of commissioning and material provision. As soon as the materials have left the warehouse, a goods issue is booked to update stock and inventory management values.

A carrier can be commissioned to deliver the materials. *Transport logistics*, in a cross-sectional function, takes on the task of booking the transport planning and the transport itself. At the end of the sales procedure, an invoice is produced for the customer. As soon as the customer has paid for the materials, payment receipt is recorded in Accounting.

1.3 Structure of the Two Volumes

This book represents the first of two volumes on core logistics operations with SAP. The chapters are arranged according to the various logistics functional areas and include the following topics:

After this introduction (Chap. 1), we turn to "SAP Business Suite" in Chap. 2, which provides an overview of the suite's components and systems as well as of the SAP NetWeaver. The aim is to make you familiar with the terminology of SAP and its components.

In further chapters, we discuss the SAP logistics components and their functions in more detail, occasionally referring to this overview chapter in order to illustrate the functions in their overall context.

Chapter 3, "Organizational Structures and Master Data", details the significance, application, distribution and links of master data for the logistics components of SAP Business Suite and its organizational structures. The organizational structure mirrors the legal and organizational setup of a company, and forms the basis of data organization in SAP Business Suite.

Chapter 4, "Procurement Logistics", primarily deals with the external procurement of raw materials and supplies. In addition to the purchasing processes involved in external procurement, we also discuss internal procurement via stock transfer. After a general overview of procurement logistics and its business-related significance, we also illustrate purchasing-specific master data and organizational structures. Subsequently, we turn to applications for requirements determination, order processing and delivery, where the ordering procedure is demonstrated using a process example. External procurement generally ends with goods receipt and the receipt of the vendor invoice. We will explain the effects of goods receipt, goods movement, integration into inventory management and the allocation of consumable materials. We then examine invoice verification and demonstrate purchasing optimization options.

Production logistics, as a link in the logistics chain, generally denotes the planning, control and interior transport of the raw materials and supplies needed for production, as well as the finished products resulting from production. In Chap. 5, "Production Logistics", we especially look at the tasks and processes of production logistics from the deployment side, as a basis of the subsequent external acquisition within the context of procurement logistics. The basics from the SAP process point of view are detailed, followed by the functional aspects of sales and procurement planning using SAP ERP and SAP APO (SAP Advanced Planning & Optimization). However, production control and capacity planning with regard to a company's own manufacturing control systems are not within the scope of this book.

Distribution logistics combines the production logistics of a firm with the procurement logistics of a customer, thus encompassing all activities related to processing orders, delivery and invoicing of the requested products. Delivery occurs from the production process or from the warehouse logistics stock. For the purposes of this book, we differentiate between distribution logistics and actual sales activities, which are aimed at the acquisition, maintenance and development of customer contacts. From the SAP Business Suite standpoint, these tasks are performed by SAP CRM (SAP Customer Relationship Management) and their functions covered by its *Account and Contact Management*. Account and contact management, as well as the maintenance of opportunities and sales activities, are thus not covered in this book.

In addition to the fundamentals of organizational and master data necessary for distribution, Chap. 6, "Distribution Logistics", also deals with order processing in SAP ERP and SAP CRM, dispatch handling and, finally, invoicing. We not only discuss pure sales, including inquiries, quotations, and order and contract processing, but also look at backorder and delivery processing and explain special distribution situations. These include returns processing as well as cash sales, returnable packaging and consignment processing.

At the end of the book, you will find a glossary, a bibliography and a detailed index that can help you find important terms and their definitions quickly.

The second volume of this series is entitled *Inventory Management, Warehousing, Transportation and Compliance*. Due to its logistic significance and related SAP applications, we have dedicated the entire second chapter following the introduction to "Transport Logistics" (Chapter 2), which illustrates the various SAP solutions regarding the topic of transport. We not only shed light on the perspective of shipping agents from the realms of manufacture and trade, but also that of transport service providers. In addition to the basics of transport logistics and its business significance and transport from the standpoint of shippers and logistics service providers, we explain in detail the individual systems and applications, and their integration into the procurement and distribution logistics systems as well as the necessary master data.

Chapter 3, "Warehouse Logistics and Inventory Management", describes warehouse logistics as a link between the internal and external logistics systems. Thus, we present those SAP processes in the realm of inventory management, goods movement and warehouse management. In doing so, we make a clear, systematic distinction between stock and warehouse management. This chapter also discusses warehouse management using the Warehouse Management solution in SAP ERP (WM) as well as in SAP SCM, SAP Extended Warehouse Management (SAP EWM).

Besides the application-specific description of the basic warehouse processes in the areas of goods receipt and goods issue, we also have a look at the fundamentals of inventory management, its evaluation and the integration of system components. We illustrate special stock and special procurement forms, consignment and sub-contract order processing based on their significance to central logistics, as well as the technical differences between WM (ERP) and EWM (SCM) processes.

Chapter 4 focuses on "Trade Formalities—Governance, Risk, Compliance". This chapter offers an overview of the functions of foreign trade and customs processing with SAP ERP and SAP BusinessObjects Global Trade Services.

Logistic control and the related reporting process are the focus of Chapter 5, "Controlling and Reporting", which also covers integration into the SAP logistics processes. We primarily examine SAP Event Management as a *Track & Trace* system for the tracking of shipments and events, the classic SAP ERP-based functions in the realm of distribution and logistics information systems, as well as SAP NetWeaver Business Warehouse (SAP NetWeaver BW). The classic reporting functions are complemented by SAP BusinessObjects. SAP thus offers the necessary tools to support users in the generation, formatting and distribution of conclusive reports, or so-called *dashboards*. Dashboards enable more than simple data evaluation, focusing on the integration and generation of intuitive visualizations that immediately display where there is a need for action.

1.4 Thanks

This book was created with the aid and the direct and indirect expertise of several SAP colleagues, whom we wish to sincerely thank.

We are very grateful to Dorothea Glaunsinger and Hermann Engesser at Springer Verlag for their guidance and support. Thanks also to Frank Paschen and Patricia Kremer at SAP-Press for their first-rate assistance during the making of the original German publication of this book. We also wish to thank translator Andrea Adelung and copy editor George Hutti.

We would like to express our special thanks to our wives and families:

- Susanne Kappauf with Leni and Anni
- Yumi Kawahara with Kai and Yuki
- Susanne Koch with David and Leah

You are the ones who, through your patience and willingness to do without a great deal of things, have made this book possible.

Chapter 2
SAP Business Suite

SAP Business Suite is composed of various business software applications. They enable companies to plan and execute processes while saving on operating costs and, at the same time, tapping into new business opportunities. The SAP Business Suite applications are based on the SAP NetWeaver platform and support the best-practice methods of all branches. In addition, the suite provides integrated business applications and functions from the realms of finance, controlling, human resources, assets management, production, purchasing, product development, marketing, distribution, service, supply chain management and IT management.

2.1 SAP Business Suite as Standard Software

The term *standard software* refers to a group of programs that can be used to process and resolve a series of similar tasks. These programs can generally be adapted to fit user-specific needs through targeted configuration, or *customizing*. Configuration in this sense means that process steps, process chains and individual functions—as reusable parts of process steps—can be influenced by entering configuration data (usually in several database tables).

Standard software can aid in preventing normal processes, such as invoice verification or transport planning, from having to be covered by the functions of individual programs. Individual software often leads to fragmented systems that are difficult to maintain, and also causes a great deal of integration work and complex data exchange efforts. Examples of standard software include the Microsoft Office package, which contains programs for word processing, address administration, presentation generation, etc., and SAP Business Suite, with its functional components mentioned at the start of this chapter.

Standard software can frequently be implemented in businesses at a much lower price than individual software. The complex, bug-ridden and pricey software development process has already been taken care of by standard software producers, such that the respective programs can be directly procured and installed. The

J. Kappauf et al., *Logistic Core Operations with SAP*,
DOI 10.1007/978-3-642-18204-4_2, © Springer-Verlag Berlin Heidelberg 2011

implementation project for new software can be directly targeted at configuring processes, generating master files, training users and transitioning from any previously used software. Due to the fact that many users are already working with the same standard software, a great deal of experience has flowed into its development and optimization, from which new customer's profit. In addition, standard solutions are regularly updated and often offer 24/7 maintenance service.

A company's decision to implement standard or individual software in a new information system must be well thought out, and depends on a few prerequisites. Standard software is flexible and frequently offers several advantages, such as a long-term guarantee of functional improvements by the producer or adaptability to changing business processes. On the other hand, the disadvantage is that the standard software might only contain 50–80% of the desired functions, with the rest having to be added via other products or individual extensions.

Whether or not these advantages apply in individual cases depends on the technical and organizational prerequisites: One criterion is whether the IT infrastructure of a company is outdated or no longer maintainable. An outdated system can often be replaced in the process of introducing standard software. Usually, however, parts of the IT infrastructure must continue to run and be linked to a new solution. In such cases, it could be that the standard software offers no suitable interface, such that either elaborate supplementary programs are required or the use of individual solutions becomes necessary. From an organizational standpoint, the standard software should at least cover the majority of a company's needs.

A further, often very important criterion is the willingness of a company to invest in a uniform software platform with the aim of gaining cost and infrastructure advantages with a harmonious IT system throughout all corporate and application areas.

2.2   Layer Models and Components of an SAP System

An SAP system such as SAP Business Suite can be divided into three layers:

* Technical hardware layer
* Technical software layer
* Application component layer

In this chapter, we will present the two primary software layers of the SAP system that bring to life business processes within a company—with a special eye to logistics.

* **The application platform SAP NetWeaver**
  This layer includes the central system (kernel) and the application server (AS), which are necessary to operate all SAP applications. SAP NetWeaver is described in more detail in the Sect. 2.3.

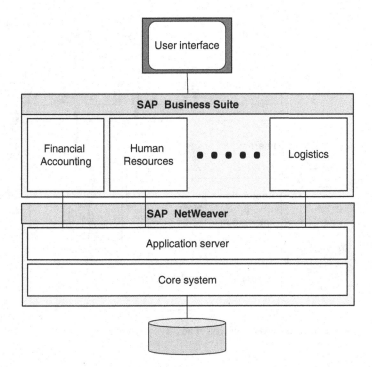

Fig. 2.1 Basic software layers of an SAP system

- **The application family SAP Business Suite**
  This layer offers a portfolio of business applications, an overview of which is provided Sect. 2.4. More detail on the logistics components can be found further on in the book.

Figure 2.1 depicts the basic software layers of an SAP system.

The software layers of an SAP system, in turn, operate on various technical hardware layers. The hardware and server layers are composed of the following elements:

- **Database server**
  The server on which the database of the SAP system is operated.
- **Application and integration server**
  The server on which the applications run and through which they are integrated.
- **Internet server**
  Servers that supply Web access to the applications.
- **Presentation layer**
  Generally the computer or hand-held device on which the consumer uses the applications.

Figure 2.2 shows how the individual software layers can be distributed over several servers (computers) with dedicated functional areas.

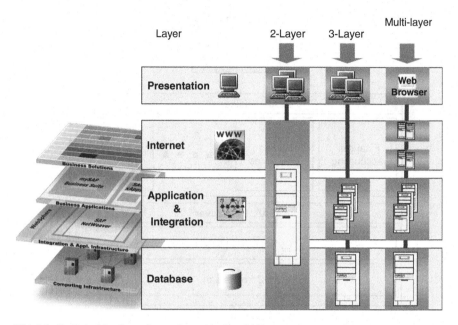

Fig. 2.2 Technical hardware layers (servers) of an SAP system

In the simplest case, only a two-layer arrangement is utilized, in which the database and application programs operate on *one* server. Access is then granted via local computers (PCs), which as a rule are equipped with a local user interface, the SAP GUI (Graphical User Interface). Optionally, the system can also be installed on a Web server to enable Web-based user interface technologies (so-called Web Dynpros).

The more common arrangement is either the triple- or multi-layer model. In the case of the triple-layer model, user access is also enabled via the SAP GUI or Web Dynpro; however, the individual application processes can be distributed over several servers within an application layer, which makes the system very easy to scale, i.e. adapt to larger numbers of users by connecting additional application servers. This scaling option also exists in the multi-layer model, which has the advantage of an additional scaling option for Internet access via a Web server.

2.3   SAP NetWeaver

With SAP NetWeaver, SAP offers a comprehensive application server to operate SAP applications and other supplementary partner products. SAP NetWeaver has a variety of functional areas and modules which together support the operation of the actual business applications. Figure 2.3 provides an overview of these areas:

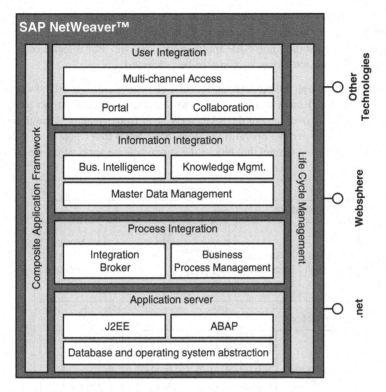

Fig. 2.3 Functional view of SAP NetWeaver

- The *application server* (SAP NetWeaver Application Server [SAP NetWeaver AS]), with its Java and ABAP stack, along with the technical database and operating system abstraction, form the basis for the technical operation of the applications.
- *Process integration* (SAP NetWeaver Process Integration, [SAP NetWeaver PI]) employs its integration broker to aid in integrating the individual applications with each other as well as other in-house systems or external business partners. The Business Process Management component represents the coordinated unit for complex business processes with preset communication procedures.
- *Information integration*, via SAP NetWeaver Master Data Management (SAP NetWeaver MDM), ensures the consistent distribution of master data in an application bundle and for uniform quality of the distributed data.
- *Knowledge management* enables the assessment of unstructured information from a variety of data sources, such as text documents, presentations or HTML files, with the aid of central, role-specific access points. It also supplies a multi-application, full-text search.
- *SAP NetWeaver Business Warehouse* (SAP NetWeaver BW) serves as the central collection of performance data from all SAP applications and other sources that can, in turn, be used for statistical evaluations.

- *User integration* provides users of application components with a central access point and a homogenous depiction of the application context from various applications (including non-SAP applications). Via the user integration and the portal, collaboration scenarios can also be processed.

Following this rough functional overview of SAP NetWeaver, we will now proceed down one level and have a look at the SAP NetWeaver basement, so to speak. Figure 2.4 shows an overview of the SAP NetWeaver from a technical standpoint.

Linked to the respective operating system, SAP NetWeaver possesses a kernel similar to an operating system that blocks the superordinate application server from

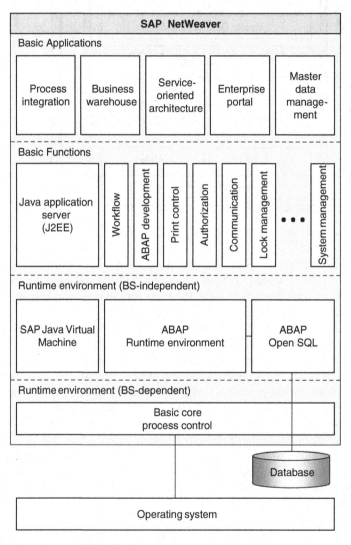

Fig. 2.4 Technical view of SAP NetWeaver

the quirks of the hardware and operating system, thus harmonizing operation of the application server.

One important advantage to the harmonization of the IT system landscape is an improved maintainability of the applications, which no longer need to worry about the respective release-specific "peculiarities" of the operating system and database.

With the ABAP and Java runtime environments, SAP NetWeaver serves as the basis of almost all SAP applications, which can thus be run in a homogenous environment. ABAP is SAP's own programming language, which began as a reporting language similar to COBOL. ABAP has been further developed into a high-performance, object-oriented programming tool that is especially suited for the development of business process applications due to its language constructs. An efficient runtime control enables independence of all application processes, such that a faulty program will have no effect on other programs.

Advantages of ABAP
Many programmers turn up their noses at ABAP, yet the fact that SAP systems are built on SAP's own programming language conveys several advantages: On one hand, SAP, and thus the user, has complete control over the extent of the language and implementation, which guarantees a high degree of quality. On the other hand, the source text of the complete ABAP-based application and base system is provided along with the software, such that your company has access to all of the possibilities of expanding it to your needs. Program code enhancements can be easily incorporated in a controlled manner via so-called *enhancement spots*, without the danger of collision with SAP maintenance releases and corrections.

In Java applications, the main complaint for a long time was a lack of process decoupling of Java Enterprise applications, i.e. if a program ran uncontrolled, it could pull down the entire application server. SAP worked on this problem as well, generating its own runtime environment for Java, the Java Virtual Machine, which provides for the independence of Java processes as they occur in the ABAP environment.

Certification According to Java EE 1.5
Version 7.1 of SAP NetWeaver Application Server Java is completely certified according to Java EE 1.5.

2.3.1   Application Support via SAP NetWeaver

SAP NetWeaver attaches AS (application support) functions to the OS-independent ABAP and Java runtime environments, which provide uniform basic functions for the entire Business Suite (see Fig. 2.4):

- **Transaction concept**
  The transaction concept in the SAP system makes sure that business operations belonging together are also consistently saved together. If, for example, an invoice is generated from a customer order, not only should the complete invoice be saved, but the order itself should consistently show the respective status.
- **Process control, work processes, load distribution**
  The process control distributes processor capacity to the individual users running application programs on an application server. In addition, it makes sure that users who have just logged on and newly invoked processes start on the server with the lowest load.
- **Lock concept**
  The lock concept prevents more than one person from making changes to a single business object at the same time (without knowledge of each other's activity). For instance, this function would prevent an employee in Dallas from editing a customer order while another employee in Portland is canceling that same order. If someone is working on an object in edit mode, other users who attempt to access the process receive a message and can only access it in view mode. In some newer transactions, a further concept with so-called *optimistic locks* is employed, which allows more than one person to access certain parts of the object.
- **User management and authorization control**
  Via a comprehensive role and authorization concept, you can precisely define which users or user groups can display, edit or run certain objects and processes. A call center employee, for example, is allowed to generate new orders and invoices but not to correct invoices because this is not part of that person's sphere of authority.
- **Print management**
  Printers can be centrally managed within an SAP system. Printer queues are provided and enable efficient processing of printer jobs from various applications. The print management also includes the integration of Adobe (Interactive) Forms.
- **Communication and service-oriented architecture**
  Technical communication within the system via application interfaces (application-to-application, or A2A) is available here, as is communication with business partners via standard interfaces (business-to-business, or B2B).

  A service-oriented architecture (SOA) is also available, which is part of the process integration. The majority of SAP applications are equipped with Web services, such that simple communication is possible. The services are pre-defined on the application object level; that is, they fulfill business operation functions (such as generation of a customer order). Further communication processes are provided by fax and e-mail integration.

- **Workflow**
  SAP NetWeaver comes with a complete workflow control that is used extensively in the applications. This workflow control can be configured in a variety of ways and integrated into the Office functions of the SAP system.

- **Development environment and Correction and Transport System**
  A complete ABAP and Java development environment is integrated into SAP NetWeaver, which makes it especially easy, particularly in the case of ABAP, to expand the system and implement current processes.

  The Correction and Transport System enables you to easily import completed developments and settings or corrections performed by SAP into other SAP systems and adapt them to the coding status of the respective system.

2.3.2 Three Significant Concepts of SAP Systems

Three significant components that concern the basic functions and applications of SAP NetWeaver as well as the SAP Business Suite are the client concept, the ability to establish organizations, and the adaptation of the systems to customer-specific business processes (called *Customizing* and the *Implementation Guide* in the SAP system).

The *client* in an SAP system is a concept for the comprehensive logical separation of various work areas within the system. Technically speaking, a client is the first key field of every application database table. When a user logs into an SAP system, he or she always enters a client number (000–999). For this, the user has to be defined in that particular client. After a successful login, the user only has access to the data and processes present in that client. Data of other clients is not accessible (it can neither be displayed nor edited). This enables separately operating organizations to work parallel to one another in their own client areas within a single SAP system without influencing one another (but a maximum of 150 clients per system is recommended for performance reasons). Via integration processes, data transfer can also take place between clients.

Thus, the client is a strict, organizational separating criterion for users working independently on an SAP system. Figure 2.5 displays the technical criteria *System* and *Client*.

For users working in separate organizations within the same company, further organizational layers are available that are developed in the applications:

- **Separate systems for corporate subsidiaries and/or regions**
  Such a case is characterized by complete logical and technical system separation, uniform usage, data integration via SAP NetWeaver PI, common or separate technical financial processing.
- **Separate system-internal clients for corporate subsidiaries and/or regions**
  This case is characterized by complete logical separation with common usage of technical system resources, uniform usage, data integration via SAP NetWeaver PI, and common or separate financial processing.
- **Internal company code for corporate subsidiaries and/or regions**
  This case is characterized by complete technical financial separation with a common use of technical system and logistics resources, uniform usage, and common or separate financial processing.

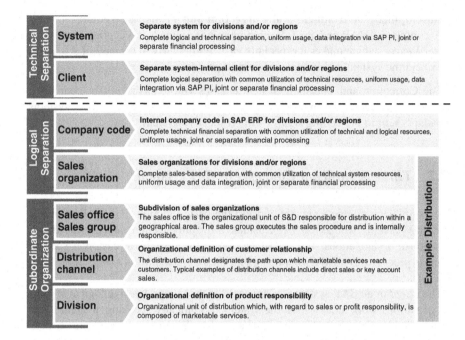

Fig. 2.5 Organizational layers of an SAP system

- **Sales organization for divisions and/or regions**
  This case is characterized by complete technical sales separation with common usage of technical system resources, uniform usage and data integration, common or separate financial processing.
- **Subdivision of sales organizations**
  The sales office is the organizational unit of the sales and distribution department, responsible for distribution within a geographical area. The sales group executes the sales procedure and is internally responsible.
- **Organizational definition of customer relationship**
  The distribution channel designates the path upon which marketable services reach customers. Typical examples of distribution channels include direct sales or key account sales.
- **Organizational definition of product responsibility**
  This case is characterized by the distribution department forming an organizational unit which, with regard to sales or profit responsibility, is composed of marketable services.

There are further specific organizational hierarchies within the individual applications. For instance, in purchasing, there are purchasing organizations and purchasing groups. In distribution, there are sales organizations, distribution channels, etc. These organizational layers enable users to work together within a single client of the SAP system, while on the other hand being limited to their respective organizational areas.

(Example: A purchaser from Chicago is not authorized to create orders for St. Louis and, of course, cannot generate an annual balance sheet in Financial Accounting.)

As explained above, configuration capability is an extremely important feature of business software. In the SAP system, this configurability is known as *Customizing* or *Implementation Guide*. The latter is divided into the individual configuration areas according to functional aspects. Using Customizing (which is really configuration and not changing the system coding), you can, for instance, very precisely control which processes within the Business Suite can be performed and how the individual process steps are composed.

Figure 2.6 shows an example from the realm of customer orders (distribution):

1. Define the available order types (sales document type), such as express order, direct sales, standard order and their functional details (such as a credit limit check).
2. Define which of your distribution organizations are allowed to use which order types (for example, only direct sales orders for shop sales).
3. Define which order items are to be permitted in which order types and what parameters are included. For example, you could exclude a service option (such as performance of maintenance) for direct sales.

Customizing an SAP system allows you to make extensive settings (in the *Sales* area alone, there are about 100 screens offering process settings), which you can use in an optimal manner where necessary to simplify processes. However, you can also deliberately keep the settings simple.

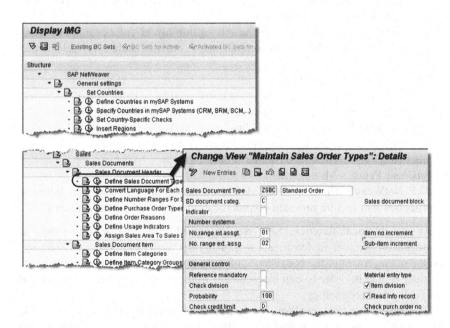

Fig. 2.6 Excerpt from Implementation guide with customizing settings for a customer order (distribution) and details of an order type

2.3.3   Application Components

SAP NetWeaver Process Integration (SAP NetWeaver PI) is responsible for data exchange (interoperability) within a system environment as well as with external business partners.

> **Definition of Interoperability**
> Interoperability is the ability of two or more system components to exchange and utilize information.

With the aid of this data exchange, you can go beyond individual applications, profiting from the functions of the integrated environment, yet still keep overall operating costs low.

SAP NetWeaver PI provides the following major interoperability functions:

- **A2A and B2B integration within and between application systems and business partners**
  Business-to-business and application-to-application scenarios can be expanded across all messaging technologies and product limits. This includes integration with Microsoft BizTalk and IBM WebSphere servers. Using communication-specific adapters, SAP NetWeaver PI can be employed for EDI communication in the various conventional formats (EDIFACT, ANSI, X.12, ODETTE, VDA, etc.).
- **Interoperability of Web services**
  SAP NetWeaver PI provides for the interoperability of corporate applications and Web services with external components, in order to enable service-oriented solutions.
- **Service-oriented architecture**
  The Enterprise Service Repository was introduced as an extension of interoperability with Web services. It contains Web service interfaces to all SAP applications. These Web services are defined on a business semantics level—that is, they possess a granularity that corresponds to access in the realm of business processing (e.g. creating a customer order).
- **Business Process Management**
  SAP NetWeaver PI enables the definition and control of complex business processes that require repeated communication between business partners.

As an additional function, SAP NetWeaver PI offers the administration of heterogeneous system environments (e.g. using the System Landscape Directory) and the management of the coexistence of various portals, whereby two portals are logically joined in order to combine content from an SAP NetWeaver portal and non-SAP portal technology.

SAP NetWeaver Business Warehouse (SAP NetWeaver BW) is a company-wide data warehouse that collects and stores data from a large variety of sources and makes it available in *InfoProvider* for the analysis of corporate performance data.

Nowadays, the integration of various source systems in a single enterprise data warehouse such as SAP NetWeaver BW poses one of the greatest challenges. Due to the sometimes extremely heterogeneous system environments, not only do various technical platforms need to be connected, but you also have to watch out for any divergent semantics of master and transaction data, in which case, by the way, it is wise to consolidate data in order to facilitate analysis. In addition, an enterprise data warehouse should provide flexible structures and layers, to enable quick response to new company developments that often occur as a result of alterations to business objectives, fusions and takeovers.

SAP NetWeaver BW covers these basic features and also offers a very high degree of performance, including the option of memory-managed data storage (via the Business Accelerator). Through integrated presentation tools (e.g. Business Explorer, BEx) and the evaluation and display functions of SAP BusinessObjects, SAP NetWeaver BW provides every user in an enterprise with an overview of required performance data precisely tailored to his or her purposes.

Employing SAP NetWeaver Business Warehouse presents several advantages:

- Less complexity, improved flexibility and very flexible data modeling
- The integration of large, complex and heterogeneous system environments with data integration across the entire enterprise
- Enables operational reporting with the aid of real-time data acquisition

SAP NetWeaver Master Data Management (SAP NetWeaver MDM) offers central master data processing within a complex integrated network. For this purpose, the software makes sure that important master data, such as business partner data or product master data, are only entered into the system once and then consistently distributed among all systems. This results in a high degree of data quality, as the data is not corrupted by duplicates or incorrect spelling. With SAP BusinessObjects tools, such as Data Services, Data Quality and Data Integrator, the laborious process of master data comparison and completion is made considerably easier. Typical master data errors that must be eliminated for an efficient and fault-free process include, for example:

- Spelling mistakes: `Gorge Miller` instead of `George Miller`
- Missing data: Palo Alto without a ZIP code
- Incorrect format: telephone number without a hyphen, `1234567` instead of `123-4567`
- Incorrect codes: currency code *CAN* instead of *CDN* for Canadian dollars
- Duplication: double master data entries, *G. Smith* and *George Smith*

SAP NetWeaver Portal provides the user-based integration of all necessary information for the respective user in a single work environment (see Fig. 2.7).

It includes access to the business processes in the SAP environment necessary to the user as well as evaluations from the Business Warehouse, Internet access to business-related content, documents from Office applications and applications from third-party providers. Access to the various systems is ensured by a single sign-on logic (SSO), in which logging into the portal once suffices, and all subsequent sign-

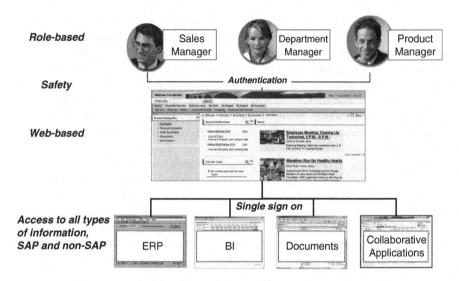

Fig. 2.7 SAP NetWeaver portal

ons are automatically authenticated via the portal. The roles defined in the portal ultimately determine which authorizations a user has in the individual systems.

SAP NetWeaver Enterprise Search—formerly known as TREX—is a cross-system search with a uniform user interface. It offers a search infrastructure necessary for universal searches and the respective data model, as well as central administration and operation functions. Many of the logistics elements mentioned in this book are already integrated into Enterprise Search, such as material, customer, sales order, supplier, delivery and order, such that you can search for these elements in a targeted manner using a free-text search.

2.4   Components of SAP Business Suite

SAP Business Suite is an arrangement of business applications based on SAP NetWeaver, designed to offer a comprehensive solution for all of a company's standardized business processes. It emerged a few years ago from the SAP R/3 system, which was released in 1992. SAP Business Suite expanded the R/3 system, having a monolithic architecture, with a series of self-contained products. Generally, complex business processes span several components of SAP Business Suite. Figure 2.8 presents an example for such a business process (make-to-order).

SAP ERP primarily consists of the components FI (Finance), CO (Controlling), MM (Materials Management), SD (Sales and Distribution), LES (Logistics Execution System), LO (Logistics), PP (Production Planning) and HCM (Human Capital Management). These components commonly form the core functionality utilized by SAP users.

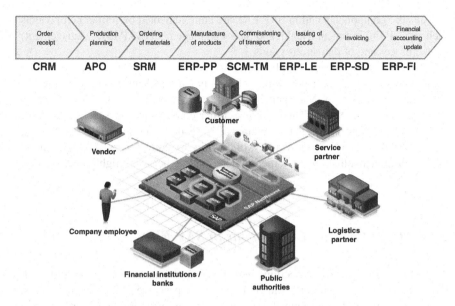

Fig. 2.8 Business process in SAP Business Suite

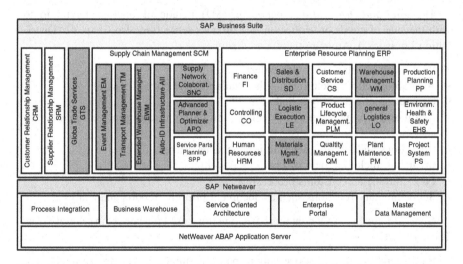

Fig. 2.9 Overview of the logistics (*gray*) and other components of SAP Business Suite

SAP ERP Central Component (SAP ECC)
The central components of ERP (FI, CO, MM, SD, LES, LO, PP and HCM) are also anchored in the SAP portfolio under the name SAP ERP Central Component (SAP ECC).

The core components within SAP ERP are complemented by further components such as SAP Environment, Health and Safety Management (SAP EHS Management) for the processing of dangerous goods.

SAP ERP has also been supplemented in the Business Suite with a number of components especially for the realms of Supply Chain Management (SAP SCM), Customer Relationship Management (SAP CRM) and Supplier Relationship Management (SAP SRM). There are also other components for special purposes, such as SAP BusinessObjects Governance, Risk and Compliance (GRC) solutions.

In Fig. 2.9, you can see an overview of the primary components of SAP Business Suite. Components that are explained in detail in this book are highlighted in gray.

2.4.1   Core Logistics Components in SAP ERP

As explained in the previous section, there are several components within SAP ERP that are directly or indirectly linked to logistics processes. These components, which are only briefly introduced here, are presented in more detail further on in the two volumes of this book.

The *Sales and Distribution components* (SD) include all functions having to do with the sale of goods or services in the broader sense (see also Chap. 6, "Distribution Logistics"). Among others, they include:

- Processing of offers and sales orders
- Availability check (determining the availability of goods for sale)
- Generation of a delivery schedule
- Credit and risk management in connection with the sales process (credit limit check)
- Conditions and pricing (determining the sales price)
- Invoicing, including payment card processing
- Foreign trade and customs processing
- Supplementary functions such as document printing, reports and analyses

General Logistics (LO) contains basic functions used repeatedly in several areas. These include:

- Batch handling, i.e. the processing of partial volumes, or batches, of a material, which are kept in stock separately from other batches of the same material (for example, a production lot)
- Handling unit management, i.e. the processing of transport containers
- Variant configuration, i.e. the description of complex products that can exist in several variations (such as cars)

The *shipping component* (Logistics Execution System, LES) primarily consists of all functions having to do with warehouse storage, shipping processing and the transport of goods (see also Chapter 6 and Volume II):

- Goods receipt process for delivered goods
- Central and decentral warehouse management (Warehouse Management, WM) with the optimization of procedures in the warehouse (Task and Resource Management) and the management of yards (Yard Management)
- Shipping preparation, shipping document generation and the goods-issue process for goods to be shipped out
- Transport preparation and processing, including freight charge calculation
- Delivery management for regular delivery routes from the distribution center to various customers (Direct Store Delivery).

Materials Management (MM) focuses on products that must be managed, procured or paid (see also Chap. 4 "Procurement Logistics"):

- Purchasing functions such as the processing of orders and order purchase requisitions
- Inventory management of materials, including materials evaluation for preparing the balance sheet and material price changes
- Invoice verification for received goods and service invoices
- Taking of inventory to determine and correct stock
- Material master data management

These four logistics components—SD, LO, LES and MM—are strongly integrated within SAP ERP and thus enable effective logistics workflow.

2.4.2   Other Logistics Components in SAP ERP

In addition to the central logistics components mentioned above, we would also like to present those components that have a somewhat more indirect connection to logistics. Further in this book, we will only discuss these components in passing:

- *Production Planning and Control* (Production Planning, PP) includes rough sales and production planning, actual production planning including capacity and requirements planning, production orders, kanban processing, make-to-stock production, make-to-order production and assembly processing. It also covers production planning for process industries.
- *Maintenance* (Enterprise Asset Management, EAM) deals with technical facilities and equipment (plant, machines, vehicles, etc.) that have to be maintained on a regular basis. It supports planned as well as unplanned maintenance and repair measures. Mobile scenarios with hand-held devices are also available.
- *Quality Management* (QM) includes functions for quality planning, testing and control, in addition to auditing management, the generation and maintenance of quality certificates and the maintenance of test equipment. In the process, you can manage inspection lots, record test results and errors, and perform sampling management tasks.

- *Product Lifecycle Management* (PLM) offers all functions necessary for product planning, design and its engineering tasks, product data distribution, the processing of product and recipe data and the respective audits.
- With the *Project System* (PS), you can generate and administer projects. This component allows you to manage project flow with its task structures and scheduling, as well as project resources and costs.
- The components for the realm of *Environment, Health and Safety Management* (SAP EHS Management) provide a variety of functions within the context of safety. Not only are product safety and hazardous materials management and testing covered, but also waste management, job safety and industrial medicine.
- The components for *Customer Service* (CS) enable you to establish a customer interface with an interaction center for other ERP components (such as Sales and Distribution, Service, and Maintenance) in order to facilitate the execution of individual processes in cooperation with customer service representatives and customers themselves.
- *SAP for Retail* has a wide scope of retail-specific functions including, for instance, point of sale or direct store delivery. In many cases, the retail functions are tailored to a very high turnover, since that is where a large number of transactions in the sales and end-consumer area are carried out.

All of the components cited enable either an expansion of logistics processes into neighboring application areas or a branch-specific shaping of logistics processes.

2.4.3   SAP Supply Chain Management (SAP SCM)

SAP Supply Chain Management (SAP SCM) complements SAP ERP with important components offering planning-based as well as execution-based functions for logistics processes. The components mentioned only briefly here will be examined in more detail in a subsequent chapter:

- *SAP Extended Warehouse Management* (SAP EWM) is the functionally very extensive successor to the SAP ERP component *Warehouse Management* (WM). It can be employed as a stand-alone system for complete warehouse management, including all contiguous processes (see also Volume II, Chapter 3, "Warehouse Logistics and Inventory Management").
- *SAP Transportation Management* (SAP TM) offers complete transportation processing, from order acceptance, transportation planning and sub-contracting, to invoicing customers and service providers. It can be operated as a stand-alone system and was also conceived for use by logistics service providers (see also Volume II, Chapter 2, "Transport Logistics").
- *SAP Event Management* is a tool with which processes can be tracked in several ways (such as transport tracking) and critical conditions in a process can be

actively determined and reported to users. SAP Event Management can be configured and used for all status management and tracking tasks (see also Volume II, Chapter 5, "Controlling and Reporting").

• *SAP Auto-ID Infrastructure* (SAP AII) integrates RFID technology into business processes. It allows users to establish a bridge between RFID readers and business processes in the application (see also Volume II, Chapter 5).

Another important component is *SAP Advanced Planning & Optimization* (SAP APO). It offers a series of functional areas that focus on long- or medium-term as well as operative planning in an enterprise:

• *Supply Chain Monitoring* serves to monitor the logistics chain.
• *Supply Chain Collaboration* enables collaboration with suppliers and customers.
• *Demand Planning* (DP) allows medium-term planning of requirements based on a prognosis for demand of your company's products on the market.
• *Supply Network Planning* (SNP) integrates the areas of Procurement, Production, Distribution and Transport. It thus enables tactical planning decisions and those pertaining to procurement sources based on a global model.
• *Global Availability Check (Available-to-Promise)* (Global Available-to-Promise, gATP) allows product-availability checks on a global basis. It also supports product substitutions and place-of-delivery substitutions (see also Volume II, Chapter 2).
• *Transportation Planning* (Transportation Planning and Vehicle Scheduling, TP/VS) enables optimal intermodal planning for incoming and outgoing deliveries. The actual transportation processing, however, takes place in ERP (see also Volume II, Chapter 2).

The optimization and planning functions of SAP APO can contribute greatly to improving efficiency, especially in larger companies with divided logistics.

2.4.4   SAP Customer Relationship Management (SAP CRM)

SAP Customer Relationship Management (SAP CRM) is a comprehensive solution to the management of your customer relationships. It supports all customer-oriented business divisions, from marketing to sales to service, and even customer interaction channels, such as the Interaction Center, the Internet and mobile clients.

2.4.5   SAP Supplier Relationship Management (SAP SRM)

SAP Supplier Relationship Management (SAP SRM) is a solution that enables the strategic planning and central control of relationships between a company and its suppliers. It allows very close connections between suppliers and the purchasing process of a firm, with the goal of making the procurement processes simpler and more effective. SAP SRM supports processes such as ordering, source determination, the generation of invoices and credit memos, supplier qualification and Supplier Self-Services.

2.4.6   SAP BusinessObjects Global Trade Services

SAP BusinessObjects Global Trade Services is a subcomponent of *SAP BusinessObjects Governance, Risk and Compliance (GRC) Solutions.* With SAP BusinessObjects Global Trade Services, you can automate international trade processes and manage business partners and documents while making sure that your company is in observance of continually changing international laws. You will find details on this component in Volume II, Chapter 4, "Trade Formalities— Governance, Risk, Compliance".

2.4.7   Other Non-logistics Components of SAP ERP

The following components are also part of SAP ERP and, for the sake of completeness, will only be briefly mentioned below:

- *Financing and Invoicing* (FI) includes G/L accounting, accounts payable, accounts receivable, bank accounting and assets accounting.
- *Cost Accounting* (Controlling, CO) provides components for general cost accounting, product cost accounting, and profitability and market segment analysis, among other topics.
- *Business Planning and Control* (Strategic Enterprise Management, SEM) supports such functions as business consolidation, business planning and simulation and profit center accounting.
- *Human Resources Management* (Human Capital Management, HCM) enables you to execute processes in the realms of human resources management, labor time management, payroll accounting, event management, personnel development and cost planning.

In many cases, these components serve as the financial backbone of the logistics process. For instance, FI and CO are frequently the first components implemented in an enterprise when the utilization of SAP logistics solutions is planned.

2.5   Summary

SAP Business Suite offers your enterprise a comprehensive and useful collection of business process components for the management of almost all business processes. SAP NetWeaver provides a favorable basis upon which to gain high flexibility with regard to hardware, databases, integration and user interaction.

The next chapter will introduce the organizational and master data that are employed in SAP logistics applications. From a technical standpoint, they form the data foundation of all logistics processes.

Chapter 3
Organizational Structures and Master Data

In the previous chapter, you received an overview of the various systems and components of SAP Business Suite. Before we take a closer look at the individual logistics areas, this chapter will be dedicated to explaining the organizational structures and master data in an SAP system and making you familiar with the most important terminology.

We will discuss special characteristics with regard to logistics subareas, the use of master data and organizational structures in the various processes, and the necessity of certain settings and parameters.

3.1 Organizational Structures

The individual elements of an organizational structure are used to map an enterprise in an SAP system. These organizational structures determine the operational framework in which all sequences and functions of logistics and financial processes occur. They also reflect the legal and organizational structure of a company and form the basis for data organization in SAP Business Suite by enabling a variety of perspectives with regard to the master data, depending on the functional business area.

> **Multiple Perspectives of the Master Data**
> From the perspective of procurement, for instance, relevant data includes information such as order quantity and delivery tolerances pertaining to purchasing. For distribution processes, significant information would focus on sales-specific data of the material master; this primarily includes the sales tax code and certain material master parameters that influence pricing.

J. Kappauf et al., *Logistic Core Operations with SAP*,
DOI 10.1007/978-3-642-18204-4_3, © Springer-Verlag Berlin Heidelberg 2011

Organizational units in SAP systems allow the mapping of the structure of a corporate organization. An organizational unit can represent an organizational or legal situation in an enterprise.

The following provides an overview of the most important organizational structures, their significance for the structuring of an enterprise, and their use in the various components of SAP Business Suite.

3.1.1   Client

The uppermost organizational level in all components of SAP Business Suite is the *client*. Within the SAP systems, the client represents a self-contained unit with regard to commercial law, organization and data, and at the top level, effects a strict separation of the master data. For this reason, users logging on to an SAP system must not only enter their user ID and password, but also must always indicate the client in which they wish to work.

The client thus divides the company into the following areas:

- Application data (master and transaction data)
- Client-based system settings and organizational structures
- User management

In practical terms, a client can correspond to a firm.

The other organizational structures are either directly or indirectly allocated to the client as the top organizational level. In SAP ERP, there is direct allocation, such as to a company code. In turn, several plants can be allocated to that company code, and several storage locations to each plant.

3.1.2   Company Code

Company codes are directly allocated to a specific client and form the second organizational level. Their primary purpose is to record all accounting-related events as the smallest organizational unit of external accounting. Company codes represent a self-contained accounting unit and are the basis for the generation of legal documentation such as annual balance sheets and profit and loss accounts.

A company code, as an autonomous accounting unit, is used in practical situations to map individual companies or self-contained accounting divisions within a client.

3.1.3 Plants and Storage Locations

Plants and their allocated *storage locations* represent those locations of a company in which materials are physically located. Plants are directly allocated to a client and to exactly one company code. Because they are assigned to a single company code—i.e. a self-contained accounting unit—inventory management and material valuation are always done on the plant level.

In business situations, plants generally correspond to a producing location or logical aggregation of spatially close locations in which material stock is present (see Fig. 3.1).

A plant can fulfill various functions in SAP ERP. As a so-called *maintenance plant,* it fulfills the task of logistically summarizing the spatial division of all maintenance objects in a certain area. In this regard, a plant can be divided into several locations and operational divisions. The division into locations takes into account spatial criteria, the operational divisions and responsibility for operational maintenance.

From a logistics standpoint, a plant can correspond to a procurement, storage, production or distribution location and can prepare and/or produce goods or services for sale or distribution:

- From the perspective of procurement and warehouse logistics, the plant is primarily that location in which material stock is present.
- From the production perspective, a plant represents a production location.
- Distribution logistics defines a plant as a distribution center, that is, the site from which materials are delivered and services performed.

A plant serving as a location for materials management can be logically divided into several storage locations. Storage locations enable a differentiated perspective, especially from a materials planning standpoint, offering a differentiation between the individual stock items within a plant. Generally, a plant can be divided into several storage locations. The individual storage locations may correspond to individual warehouses or areas in which stock is stored.

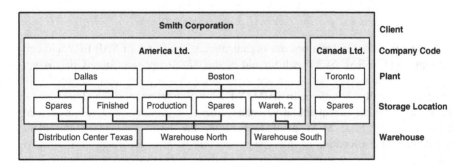

Fig. 3.1 Relationship between client, company code, plant, storage location and warehouse number

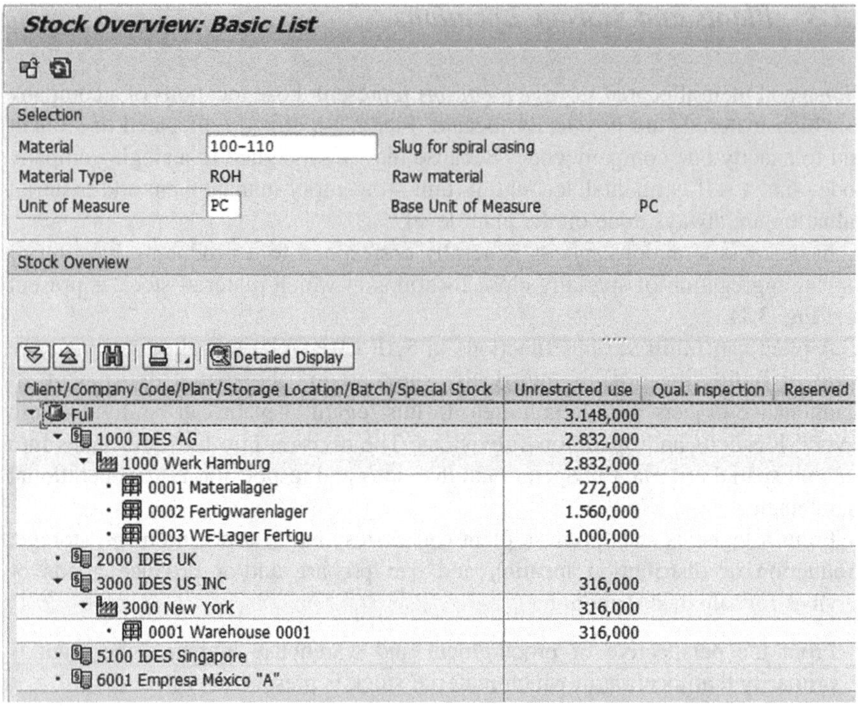

Stock Overview: Basic List

Selection

Material	100–110	Slug for spiral casing
Material Type	ROH	Raw material
Unit of Measure	PC	Base Unit of Measure	PC

Stock Overview

| Detailed Display | | | | |
Client/Company Code/Plant/Storage Location/Batch/Special Stock	Unrestricted use	Qual. inspection	Reserved
Full	3.148,000		
1000 IDES AG	2.832,000		
1000 Werk Hamburg	2.832,000		
0001 Materiallager	272,000		
0002 Fertigwarenlager	1.560,000		
0003 WE-Lager Fertigu	1.000,000		
2000 IDES UK			
3000 IDES US INC	316,000		
3000 New York	316,000		
0001 Warehouse 0001	316,000		
5100 IDES Singapore			
6001 Empresa México "A"			

Fig. 3.2 Stock overview of plant and storage location stock

Plant/Storage Location Stock Overview
Figure 3.2 shows the stock situation of Material No. 100–110. Plant 1000, divided into three storage locations, has a total stock resulting from the cumulative stock of the various storage locations.

Plants and storage locations are organizational structures of SAP ERP and can be replicated in SAP SCM with the aid of the *APO Core Interface* (CIF). When the plants and storage locations are replicated, the SCM system automatically generates so-called *locations*. In SAP SCM, a location refers to a logical or physical location in which the quantitative management of products and resources can take place. In SAP SCM plant replication, the basic settings of the plant are copied, and a location of the type *production plant* is generated in SAP SCM (see Fig. 3.3).

You will find a more detailed description of the master and organizational data replication between SAP ERP and SAP SCM in Sect. 3.3.1.

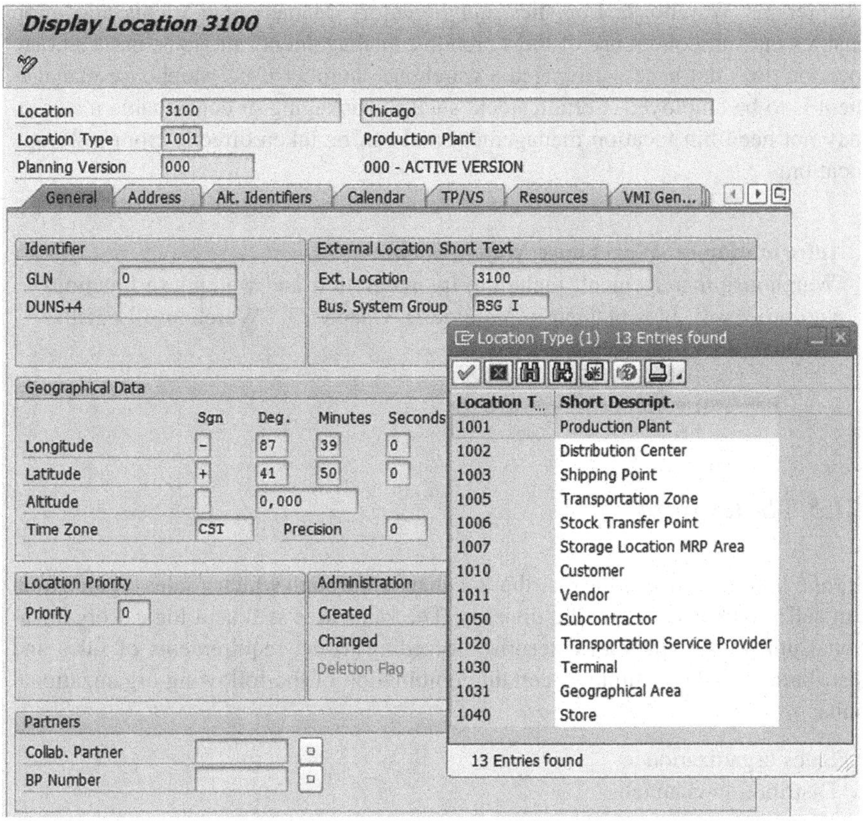

Display Location 3100

Location	3100	Chicago
Location Type	1001	Production Plant
Planning Version	000	000 - ACTIVE VERSION

General | Address | Alt. Identifiers | Calendar | TP/VS | Resources | VMI Gen...

Identifier		External Location Short Text	
GLN	0	Ext. Location	3100
DUNS+4		Bus. System Group	BSG I

Location Type (1)   13 Entries found

Location T...	Short Descript.
1001	Production Plant
1002	Distribution Center
1003	Shipping Point
1005	Transportation Zone
1006	Stock Transfer Point
1007	Storage Location MRP Area
1010	Customer
1011	Vendor
1050	Subcontractor
1020	Transportation Service Provider
1030	Terminal
1031	Geographical Area
1040	Store

13 Entries found

Geographical Data

	Sgn	Deg.	Minutes	Seconds
Longitude	-	87	39	0
Latitude	+	41	50	0
Altitude		0,000		
Time Zone	CST		Precision	0

Location Priority		Administration	
Priority	0	Created	
		Changed	
		Deletion Flag	

Partners		
Collab. Partner		
BP Number		

Fig. 3.3 Plant location in SAP SCM

3.1.4   Warehouse Number

A *warehouse number* is based on subordinate organizational units, such as warehouse type and storage areas, with which the spatial circumstances of a warehouse are depicted. The warehouse number thus represents a technical and organizational unit of a complex warehouse system. All storage-specific material master data, such as information on palletization and put-away into and removal from storage, is stored on the warehouse number level.

Warehouse numbers are allocated to a specific plant/storage location combination, and it is possible to allocate a variety of plant/storage location combinations to a single warehouse number. Simultaneous allocation to several warehouse numbers is not possible (see Fig. 3.1). The assignment of a warehouse number to a plant/ storage location combination represents a link between inventory management and warehouse management, which will be explained in more detail in Volume II,

Chapter 3, "Warehouse Logistics and Inventory Management". This allocation enables the subsequent use of the warehouse management functions, but a storage location does not necessarily need a warehouse number if no warehouse management is to be employed. Certain stock, such as packaging or consumable material, may not need bin location management and can be taken directly from a storage location.

Information on Warehouse Management
Warehouse management, including its integration and warehouse functions and processes, is explained in Volume II, Chapter 3, "Warehouse Logistics and Inventory Management".

3.1.5   Sales Area

A *sales area* determines the distribution channel through which a sales organization can sell products of a specific division. The sales area is thus a logical organizational unit that divides an enterprise according to the requirements of sales and distribution, and constitutes a certain combination of the following organizational units:

- Sales organization
- Distribution channel
- Division

Within the framework of distribution logistics, the *sales organization* is responsible for the sale and distribution of materials and services. It typically represents a selling unit in the legal sense.

In SAP ERP, the sales organization is an important feature for controlling all sales-related business activity and is obligatory in all sales documents. Using it makes it possible to regionally divide a market and its sales-related activity not only according to legal aspects but also certain criteria.

Distribution channels represent the various distribution routes via which marketable materials reach customers. They can be employed to control business activities within a sales organization based on varying prices, minimum order amounts and suppliers. A single distribution channel can be allocated to several sales organizations. A typical example for the use of distribution channels is the differentiation of wholesale, retail and direct sales. Depending on the distribution channel, this differentiation allows various prices to be determined for customers belonging to a particular sales area.

The *division* represents an organizational unit that can be used to control logistics activity for marketable materials according to operational responsibility or profit responsibility.

The sales organization is allocated to exactly one company code in which the sales processes are recorded for accounting reasons. The sales organization can be divided into various distribution channels, and several divisions can be allocated to each distribution channel. A number of plants can be assigned to a single distribution chain and at the same time belong to various company codes. If the sales organization and the plant are allocated to different company codes, an internal adjustment can be performed to facilitate the accounting of the plant's business transactions. This procedure is called *cross-company sales*.

For each sales organization, the plants authorized to sell are determined based on the distribution channel, such that a sales organization can sell goods from more than one plant. At the same time, a plant can be allocated to various sales organizations. All of these sales organizations can sell goods from that plant. Figure 3.4 shows the cardinality of the various organizational structures in SAP ERP.

In order to subdivide a sales area into geographical responsibilities and establish regional contact to the markets, *sales offices* can be created and allocated to one or more sales areas. Here, the sales office represents either a contact person or organizational unit, such as a branch store that customers can visit. *Sales groups* are a further subdivision of a distribution chain and are directly allocated to a sales office. They represent a group of employees and contact persons in the respective sales offices.

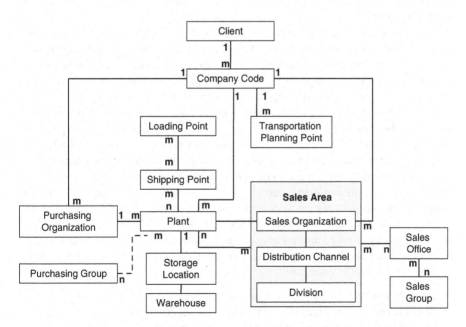

Fig. 3.4 Cardinality of the organizational structures in SAP ERP (1:m = one to many, m:n = many to many)

3.1.6 Shipping Point

The *shipping point* is an organizational unit of distribution logistics, and represents a place or group of people responsible for shipping activity. The responsibility, and thus the processing loading point, can be based on the following criteria:

- Delivering plant
- Dispatch type (rail, truck, airplane, etc.)
- Necessary loading equipment (forklift, elevating truck, etc.)

Actual shipment is processed by a single shipping point. The shipping point is allocated to one or more plants and, in turn, can be subdivided into so-called *loading points*. The loading points can correspond to such things as individual ramps via which loading onto a truck takes place.

3.1.7 Purchasing Organization and Purchasing Group

The *purchasing organization* is an organizational unit used to organize an enterprise according to purchasing demands. The actual procurement of materials and services always takes place in relation to a purchasing organization. For this reason, conditions worked out with suppliers as well as purchasing-specific master data are maintained here.

The allocation of a purchasing organization to a company code is generally optional. If allocation is chosen, a purchasing organization can only be assigned *one* company code. Procurement within the framework of procurement logistics can thus be done on a companywide basis, meaning that a purchasing organization can supply all company codes of a corporation or specific company codes.

Allocation of a plant to a purchasing organization is also optional. If this is done, a plant can be allocated to more than one purchasing organization. It is also possible to assign a *standard purchasing organization* to a plant. If several purchasing organizations are assigned to a single plant, the standard purchasing organization controls source determination for consignment and stock transfer orders.

In order to minimize maintenance efforts, especially in the realm of contract management and with regard to generating condition records for pricing, a *reference purchasing organization* can be allocated to a purchasing organization. The purchasing conditions of a reference purchasing organization and its contracts can then also be utilized by other purchasing organizations. Thus, the following enterprise scenarios can be configured:

- A single purchasing organization procures for one plant.
- A single purchasing organization procures for several plants.
- Several purchasing organizations procure for one plant.

A *purchasing group* is a purchaser or group of purchasers. Purchasing groups generally tend to the acquisition of certain materials or services or, depending on purchasing area, serve as contacts for external and internal suppliers. Their duties in this regard include operative planning and order processing as well as the maintenance of purchasing-specific master data, especially pertaining to purchasing conditions. In contrast to purchasing organizations, purchasing groups neither control the procurement procedure nor are on the data management level. Instead, they serve as selection criteria and a level for evaluation in the information system as well as for the assignment of authorizations. Allocation of purchasing groups is thus done directly via the procurement manager or purchasing agent who is directly assigned to a purchasing group.

3.2   Master Data

Transaction data is variable, generally time-limited and used by certain applications. It is mainly created in an SAP system based on business transactions and processed as so-called *documents* in its business context. Examples for documents and thus for master data include purchase orders, contracts and sales orders.

Master data, on the other hand, refers to data records that remain unchanged over an extended period of time and are stored in the database for lengthy periods. Master data includes customer master data records, material master data, and conditions in procurement and sales and distribution.

In order to prevent redundant data maintenance, master data is used within the SAP system in a cross-application manner and, if required, exchanged between the systems of SAP Business Suite. The advantage of this integration is that the time needed to process business transactions is reduced considerably, because the master data is automatically integrated into the processes and need not be updated redundantly.

Master data is utilized by all corporate departments. The delineation of these departments and thus the maintenance of their respective master data are done analogously to the organizational structures and corporate responsibilities. The following master data types are especially significant to logistics:

- Business partners such as customers and vendors
- Material masters
- Prices and conditions

In this section, we will therefore present a thorough overview of the most important, central master data in an SAP system, its significance for logistical processes and, if applicable, its integration and distribution. The subsequent chapters will highlight the special features of this and other master data in the various logistical processes.

3.2.1 Business Partners

Business partners are all legal entities or individuals with whom a company maintains business contacts. In this regard, we can basically differentiate between customers and vendors. From an accounting point of view, all customers with whom a company is in contact are *debitors*. Suppliers, or vendors, by whom deliveries or services are provided, are called *creditors*. A business partner can be a debitor and a creditor at the same time.

Customers are generally business partners to whom goods and services are sold within the framework of distribution logistics. Depending on its relevance to distribution or accounting, customer master data is maintained in various views (more information on views is available in Chap. 6, "Distribution Logistics"). In an ERP system, the view corresponds to the organizational divisions of a company and thus the level and functional area in which master data is utilized.

Basically, there are three levels, or views, involved in the maintenance of a customer master record:

- General data
- Sales area data
- Company code data

General data is valid for all organizational units and is thus independent of the accounting or sales-specific structure of a company. In addition to names, addresses and contact information, general data also includes control data, export data, marketing information and information regarding payment procedures. Control data refers to master data pertaining to account control, customer location-specific transport zones and further information, such as tax code numbers or tax jurisdictions. Payment transaction data primarily includes customer bank account data. As a rule, the general data is entered by the department that also creates the master record for a business partner.

Sales area data includes all sales-related information for a specific customer. Maintenance of the sales area data, a certain combination of sales organization, distribution channel and division, is a prerequisite to the collection of sales activity data for a customer. The sales data thus includes data regarding pricing, invoicing and terms of delivery and payment, as well as delivery priorities and shipping conditions (for sales area data of a debitor, please see Chap. 6, "Distribution Logistics").

Similar to the sales area data, the company code data only applies to one specific company code for which data pertaining to a particular customer is maintained. Company code data is accounting data, and enables a debitor-based view of the customer. For this reason, such data contains accounting and company-code-specific information regarding the account management of the debitor, payment transactions and settings for the dunning procedure and correspondence.

When a customer is maintained in the system, a specific number is issued for the debitor. The type of number issued, whether it is an external number issued by an administrator or an internal number via the system, depends on the *account group* of the customer. The customer number also functions as the subledger number for financial accounting. The sum of receivables per customer is entered and updated for subledger accounting. Because these figures are accounting receivables of the company particularly stemming from sales activity, a *reconciliation account* must also be maintained. The reconciliation account is a company-code-specific master file in the customer master record and corresponds to a G/L account in general ledger accounting. In G/L accounting, it represents the receivables of a company toward customers.

Due to direct communication with the customer via the various communication channels, such as call centers, Internet portals and e-commerce, an SAP Customer Relationship Management System (SAP CRM) will generally be the leading system for the maintenance and generation of business partners. Business partner data is seamlessly exchanged with SAP ERP, where it corresponds to a customer master record. Data exchange (also: replication) takes place via the *CRM Middleware*, which is explained in Sect. 3.3.2.

In addition to general data, the customer record in SAP CRM contains information from the marketing and service views. Such information includes marketing attributes, business partner relationships and a complete history of all activity (such as a transaction and interaction history) that has taken place regarding a customer.

The data of a business partner in SAP CRM is maintained in accordance with the *business partner function* to which a business partner is allocated. These roles correspond to the business transactions in which a particular business partner can be involved, and serve to classify a business partner in operational terms and maintain that partner's master data in a role-specific manner. For example, less master data must be maintained for the role of "Prospect" than for the role of "Customer". In this context, a business partner can take on more than one role. Figure 3.5 shows a customer master record in SAP CRM.

Fig. 3.5 Customer master record in SAP CRM

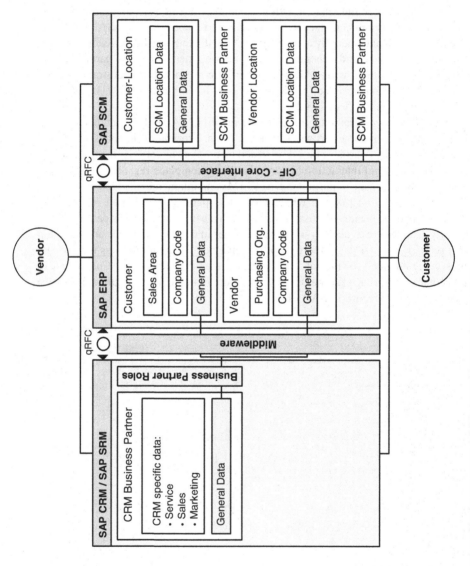

Fig. 3.6 Integration of business partners (qRFC: queued Remote Function Call)

A customer hierarchy enables the flexible mapping of a certain customer structure. This structure often corresponds to the organizational structure of a customer or reflects a customer's process organization with regard to sales or purchasing activities. Such customer organizational structures are either mapped in SAP CRM as the *account hierarchy* or a so-called *buying center* and maintained in SAP ERP as the *customer hierarchy*. The customer hierarchy is employed during the processing of orders or invoices for partner determination and the generation of statistics.

A *vendor* is a business partner who sells goods and services to a plant or customer. For the maintenance of a vendor record, there are three levels, or views:

- General data
- Company code data
- Purchasing organization data

Due to the central importance of the vendor for procurement and source determination and of the vendor's control function with regard to purchasing, we will have a detailed look at the vendor master record in Chap. 4, "Procurement Logistics".

Figure 3.6 shows the integration of business partners. The leading system for the generation and maintenance of business partner data is usually SAP CRM. In it, the business partners are maintained with their roles, which are transferred via CRM Middleware to SAP ERP. CRM-specific data, which is exclusively used and maintained in SAP CRM, is excluded from this transfer and is not replicated. SAP ERP can generate either a customer or vendor master record from the general data, in accordance with the allocated business partner function. This master data is supplemented by purchasing or sales-specific data, and can be transferred to an SCM system via the APO Core Interface (CIF). The transfer of this master data causes a location and, optionally, a business partner, to be maintained in SAP SCM. The kind of location—that is, whether or not it is a customer or vendor location—is determined by the location type (see also the location types in Fig. 3.3).

Section 3.3, explains the function of CIF and CRM Middleware.

3.2.2   Material Master

Products and services are centralized in SAP systems under the term *material*. The *material master*, the entirety of all material master records stored in an SAP system, contains all required information regarding the materials that a firm produces, procures, stores or sells. This information includes all parameters necessary for the management of the respective material and its stock and thus its use in procurement, production, distribution and warehouse logistics.

The material master is the central source in an SAP system for obtaining material-specific information. The integration of all data in a single material master record eliminates the problem of having to keep redundant data, since all information regarding the processes of purchasing, distribution, production and warehousing, as well as data on all operative functions and corporate areas, can be used communally. For example, in distribution logistics, the material master forms the basis for the distribution process, and can be accessed by those processing inquiries, quotations and orders to obtain relevant data. The same is true for all other functional areas of a company.

With regard to logistics and thus the context of this book, the material master is especially utilized by the following functional areas:

- by *procurement logistics* for the procurement and ordering procedure and invoice verification
- by *warehouse logistics* and *inventory management* for goods movement and physical inventory management
- by *production logistics* for material requirements and procurement planning
- by *distribution logistics* for activities related to sales and distribution

The hierarchical structure of the material master in SAP ERP mirrors the organizational structure of a company. According to this grouping, certain material data is valid for all organizational levels, while other data only pertains to a particular organizational structure. This material master structure eliminates redundant storage of material data, especially in cases where the same material is used in more than one plant, sold by more than one sales division or stored in more than one warehouse location.

The uppermost level thus contains general data, or *basic data*, as it is called. This level consists of client-specific information that applies to every plant and warehouse within a corporation. Such information includes basic quantity units, weights and dimensions, as well as warehouse conditions and information on such stipulations as whether or not a material is classified as explosive or toxic.

The next level is the *plant level*, containing plant- or division-specific information, such as how procurement is to be performed, which maximum and minimum order quantities influence procurement, or the reorder level for a particular plant. Analogous to the plant level, distribution information in a material master is also dependent on the organizational structures of the Distribution department and can be maintained individually for every sales organization and distribution channel. The warehouse location and warehouse number-dependent level contains data that specifically applies to a single warehouse location or a certain warehouse.

Data in a material master is divided into individual areas, depending on its respective department, called *views* (see Fig. 3.7). Each view contains the necessary data for that particular department. To select a certain view, you usually must enter the corresponding organizational structure.

For instance, the distribution view, which contains information on sales order processing and pricing, requires the entry of the sales organization and distribution channel. The procurement view contains data entered by the Procurement department regarding a specific material. On the other hand, data for material

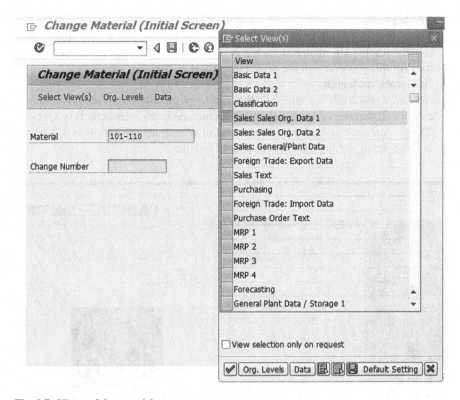

Fig. 3.7 Views of the material master

requirements planning and forecasting of material requirement are found in the planning and forecast views (see also Chap. 4, "Procurement Logistics").

In addition to logistical information, the material record contains evaluation and calculation-specific data that is maintained by the respective department in the controlling and accounting views.

The views of a material record, the maintainable fields, the order in which data screens appear and thus the operational use of the materials is determined by the *material type* to which a material record belongs.

A material type has an important control function and primarily determines the criteria governing how material accounting and inventory management are to be performed by determining which accounts are to be used to book material movement. In addition, allocation to a material type determines the number range and type of number assigned, either internal or external. Furthermore, it determines and controls the quantity and value update, as well as the procurement type of a material—that is, whether it is manufactured in-house or externally, or whether or not both types of procurement are possible.

Material types can be created individually and set according to operational requirements. In SAP ERP Standard, the following material types, among others, are available: trade goods, non-stock material and services.

- **Trade goods**
  Trade goods are movable goods that are bought by a company, stored and usually resold. The material record of these materials receives the necessary data for this procedure in the procurement and distribution views.
- **Non-stock materials**
  Non-stock materials are materials that can be physically stored but not counted as stock. Examples of non-stock materials include items such as nails or screws. The procurement and account assignment of consumable materials is described in more detail in Chap. 4, "Procurement Logistics".

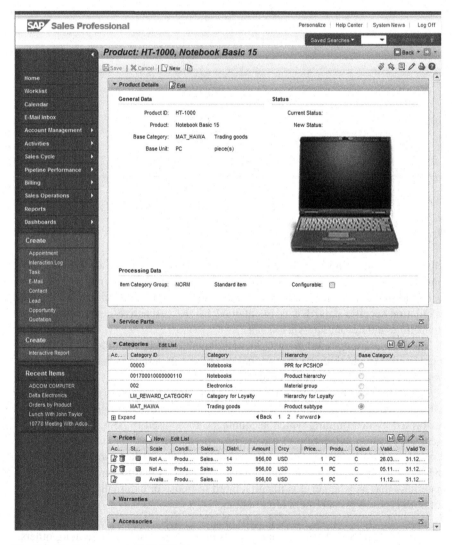

Fig. 3.8 Product master in SAP CRM

- **Services**

  Services and other *intangible goods* are depicted as material in an SAP system. These materials differ from other materials particularly due to the fact that procurement and consumption occur simultaneously. For this reason, services are neither storable nor transportable. Because services cannot be stored, a material master record of this material type does not contain warehouse or stock data.

Material master records are utilized in all systems of SAP Business Suite. The integration of the material masters is done via their distribution. Generally, SAP ERP, with its logistical core processes, represents the leading system for the maintenance of material master records. Material masters are distributed to the connected SAP systems depending on operational needs and the system environment implemented.

The material master in a CRM or SRM system is called the *product master*. Product information is used in SAP CRM for marketing, sales and service processes, while SAP SRM focuses on procurement processes. The maintenance and

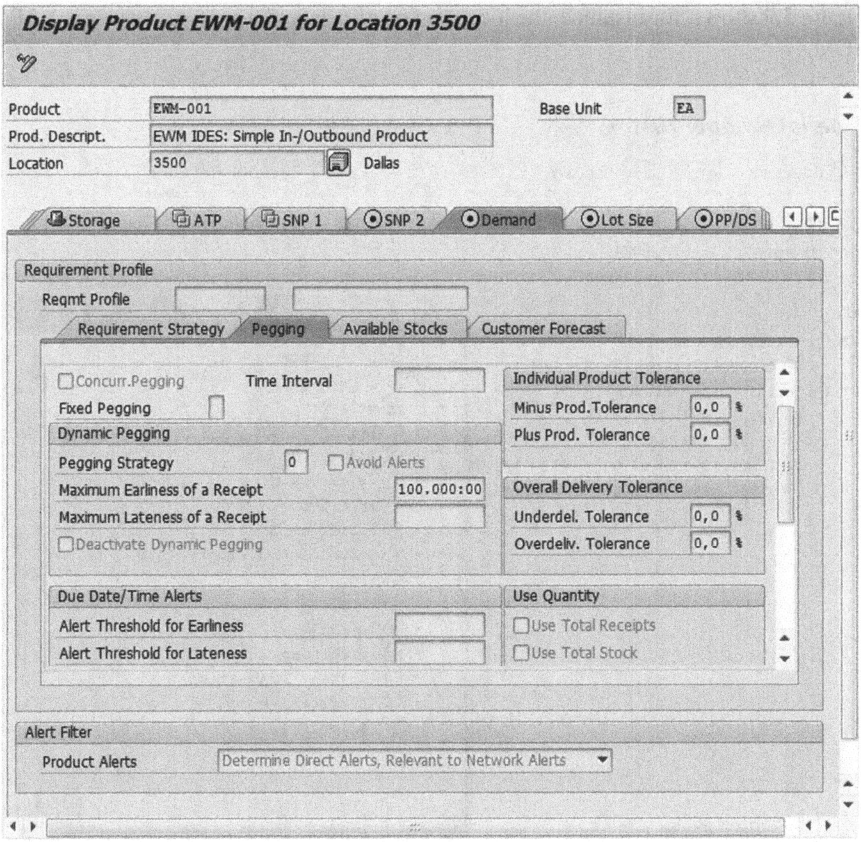

Fig. 3.9 Product master in SAP SCM

generation of products are done either directly in the systems cited or through a seamless distribution of the SAP ERP material master via CRM Middleware. Figure 3.8 shows a product master in SAP CRM.

In addition to general data and cross-system product master data, the CRM product master contains data that is used exclusively in SAP CRM. Such data includes, for instance, information on competitors' products or product-specific attachments such as images or operating instructions.

The product master in SAP SCM is usually transferred to the SCM system from ERP via the APO Core Interface (see also Fig. 3.13). As a piece of master data of the so-called SCM base, a product has access to all SAP applications of the SCM system. These primarily include SAP Advanced Planning and Optimization (SAP APO) and SAP Extended Warehouse Management (SAP EWM).

Similar to the general and plant-related material master records in SAP ERP, the SCM product master can be divided into global and location-related data. Global data in this case, analogous to the general material master data, is valid for all locations. Location-specific data includes settings regarding procurement, lot size and availability checks for a particular customer or plant location. Figure 3.9 shows, for instance, the location-specific data of a product master in SAP SCM.

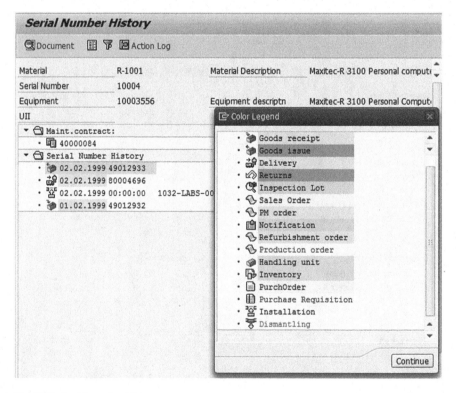

Fig. 3.10 Serial number history

Serial Number History
Figure 3.10 shows the serial number history of material R-1001 with serial number 10004. This material, a personal computer, was acquired on February 1, 1999, and delivered to and installed for a customer on February 2, 1999. The data pertaining to the customer for whom the computer was installed, the location and individual status of the technical object are stored in an equipment master record. Equipment master records belong to the technical objects, which are not explained in further detail in this book.

In accordance with the settings in the serial number profile, a serial number was automatically issued by the system on February 1, 1999, along with the procurement of the material requiring a serial number. Delivery to the customer was then done with reference to the existing serial number, whereby serial numbers were to be recorded for every material item requiring one.

A *serial number* identifies serialized material and enables individual differentiation of materials with regard to their movement history. From a maintenance and service standpoint, serial numbers are issued especially for technical objects, or *equipment,* and allow seamless tracking with the aid of the serial number history. The serial number requirement of a material, that is, the characteristic that the serial number must be indicated in addition to the material number in all logistical processes, depends on settings in the serial number profile.

The *serial number profile* is assigned in the plant view of the material master. This profile controls whether serialization can or must occur for a certain material and operative procedure, and whether it can be done automatically or not. Furthermore, this system setting defines whether a new serial number is to be issued to a particular procedure or whether an existing number should be used.

3.2.3   Batches

Batch management enables the logistical management of materials based on a homogeneous subset. *Batches* represent product quantities that can be concentrated in a homogeneous subset based on certain attributes, or *specifications.* Such a homogeneous subset, the batch, is primarily formed on the basis of legal requirements and allows seamless tracking and a differentiated inventory management with regard to quantity and value. The batch is especially useful in enabling a differentiated procedure in the realms of sales and external procurement, based on one particular product specification and attribute. In the area of production, it facilitates practicability decisions and serves as a monitoring aid for internal planning.

Batch-Managed Materials
Batch-managed materials include, for instance, drugs, foods and generally all products for which separate handling is required due to the production process, fluctuations in quality or for reasons related to shelf life.

Batches can be used in all logistical processes and are always clearly assigned to one material. The *batch management requirement*, meaning whether or not a material is to be managed in batches, depends on the settings in the material master. The batch itself is identified by a batch number, which is either issued on the plant, material or client level. These *batch levels* enable distinct identification and number assignment for all materials of a specific plant (plant level), for several plants on the material level and within a single client level. The batch number is explicitly defined for the selected batch level.

The *batch specification* indicates the technical, physical or chemical characteristics of a certain batch. Batch specifications, which are the attributes and characteristics of a batch, can be freely defined. They are based on a *Classification*, a cross-application core component of every SAP system.

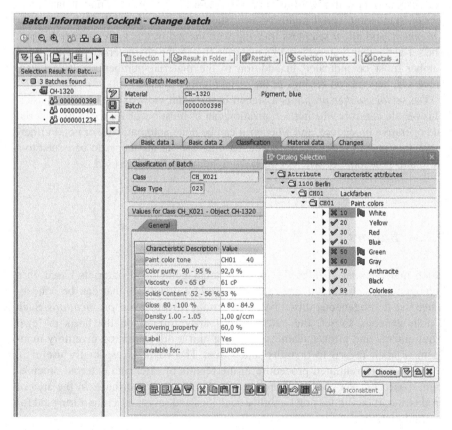

Fig. 3.11 Batch information cockpit

> **Batch Information Cockpit with Batch Classification**
> Figure 3.11 shows the Batch Information Cockpit, the central work area in
> SAP ERP for the selection and maintenance of batches. This example shows
> the batch-managed material CH-1320, for which a batch exists. Batch 1 has
> been classified, and specifies the material as a labeled, blue paint color
> approved for sale in Europe. In addition to the batch specification, the basic
> data also includes the date of manufacture and shelf life, as well as batch
> conditions and trade data.

Batch determination allows you to find a certain batch with very precise
specifications within a logistical chain. Automatic batch determination, which can
be expanded to include your own search strategies, finds batches suitable for a
certain business transaction. Actual determination is done using a batch search
procedure stored in the system, equipped with certain search strategies. These
search strategies take into account specific selection criteria and enable a targeted
search, for example according to remaining shelf life, delivery date or batch
specifications. The system checks the availability of the batches and generates a
quantity recommendation. In the distribution realm, batch determination is done
with a batch availability and applicability test for the batch found by the system
based on specifications indicated by the customer.

The *Batch Information Cockpit* is the central work area in SAP ERP for the
selection and maintenance of batches, and offers comprehensive analysis and
control options (see Fig. 3.11). In addition to the selection of batches related to a
certain attribute, for instance to determine batches whose shelf life is almost
expired, the batches found can also be selected and placed into a work list.

In addition, the cockpit offers a batch where-used list and a batch-specific
availability overview, as well as an inventory overview with batch attributes.

3.2.4   Prices and Conditions

Prices and conditions are stored in the so-called *condition records*. All prices,
discounts, additional charges, freight charges and taxes that accumulate in daily
business can be stored in the system as a condition record and thus be available
to the respective processes as a price element. During document processing, the
conditions are automatically assumed for the various logistical processes, in procurement, distribution and transport processing.

Maintaining condition records (see Fig. 3.12) is done with relation to a certain
condition type. The condition type reflects a certain aspect of the operative price
determination activity and determines the basic use of a condition. Every type
of price, discount or additional fee that can occur in the individual business
transactions is represented by its own condition type.

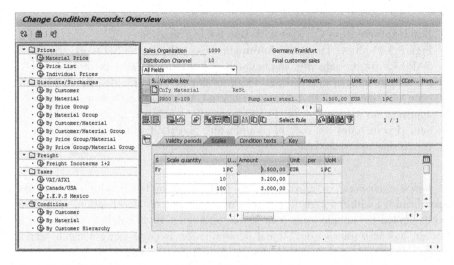

Fig. 3.12 Condition maintenance in sales and distribution

Condition Maintenance in Sales and Distribution

Figure 3.12 shows the maintenance of a condition record in Sales and Distribution. Condition type PR00 contains the sales prices of material P-109. The sales price of this material is based on the sales organization and distribution channel for which the condition record was generated. The actual condition rate depends on a scale stored in the system.

In addition to the type of condition, the condition type also controls whether it is to be categorized as a percentage, quantity-dependent or sum-dependent discount or addition. The actual condition value within a condition record can still be based on a certain scale. In order to ensure that a price agreement can be limited to a specified period, condition records are maintained for a certain period of validity.

Price determination, and thus the determination of condition values for the individual condition records, is done directly in the various documents based on a so-called *pricing procedure*. The pricing procedure is determined for a certain document type and customer, and contains all conditions necessary for a specific business transaction. Such conditions can include pricing, taxes, additional fees and rebates.

For automatic price determination, the system takes the corresponding data from the condition records and uses it to determine the amounts for the respective price elements as well as the final amounts to be paid or collected. Determination of the condition records is done according to the so-called *condition technique*. For this purpose, the document is assigned to a condition schema that contains all conditions possible for a certain business transaction. Via an access sequence, each condition controls with which criteria and in which condition tables the condition records are to be sought. For each condition of that schema, the system then automatically searches for valid condition records based on a certain combination of attributes, and establishes the condition value.

> **Information on Price Determination**
> We will discuss the topic of price determination for orders and customer orders in Chap. 4, "Procurement Logistics", and in Chap. 6, "Distribution Logistics".

3.3 Integration and Distribution

In order to guarantee seamless integration of varying SAP systems and the business transactions processed with them, certain master and transaction data are distributed among the systems. The integration and distribution of master and transaction data can synchronize objects within SAP Business Suite that are similar with regard to business aspects but different technically. These synchronized objects can be organizational structures and master and transaction data themselves.

The technical integration and exchange of this data is done from an ERP standpoint primarily via CRM Middleware and APO Core Interface (CIF). Figure 3.13 shows the system integration between an ERP and a CRM system as well as an SCM system.

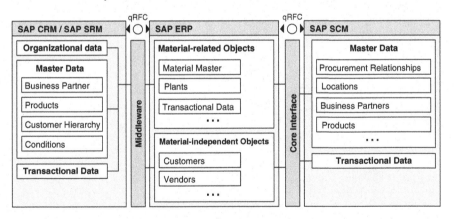

Fig. 3.13 Overview of the integration of SAP ERP, SAP CRM and SAP SCM

3.3.1 APO Core Interface

The *APO Core Interface* (CIF) is a real-time interface for the integration of SAP ERP with SAP SCM. The CIF enables the provision of initial data as well as the supply of SCM-based systems with data alterations.

The system connection between the two systems is achieved via a so-called RFC connection (*Remote Function Call*). The special feature of this type of communication between the systems lies in the asynchronous processing of the data transfer. This means that the data from the sending (ERP) system is first buffered and then transferred, or that it is transferred and subsequently buffered by the receiving (SCM) system, then processed.

Outbound and inbound processing is performed in sequence in an outbound or inbound queue. In the event of an error, caused for instance by a failed network connection, this queue saves all transfers and enables seamless continued

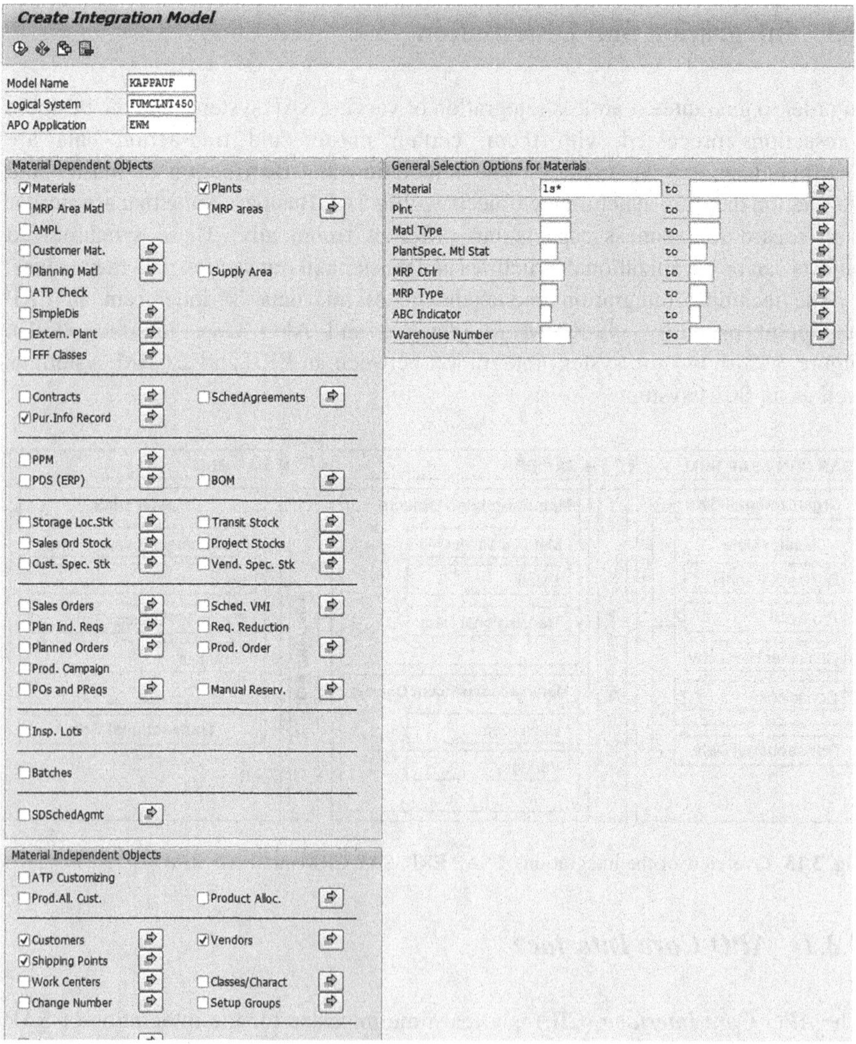

Fig. 3.14 Integration model for the communication of SAP ERP with SAP SCM

processing after the error has been located and eliminated. The *queue* is a type of waiting line that enables real-time exchange and processing of information, making possible, for example, SCM-based planning in real time. This type of RFC invocation is known as *queued Remote Function Call,* or *qRFC* for short.

The transfer of master data requires the creation of an integration model (see Fig. 3.14).

Integration models contain the necessary parameters indicating which master data in the ERP system is to be selected and transferred. The model is activated after its generation. This means that data to be transferred to SAP SCM is selected via the integration model. The data involved are generally either material-related or material-independent objects. *Material-related* objects include in particular materials and plants, contracts, delivery plans and procurement information records necessary for procurement, as well as customer orders and planned independent requirements as the basis for requirement determination in SAP SCM. *Material-independent objects* include shipping points, and customer and vendor master data. Customers and vendors are mapped as locations in SAP SCM, similarly to the plants in SAP ERP (see Fig. 3.3). The locations can then be supplemented with SCM-specific data.

One-Way Transfer
The transfer of master data is only performed in one direction, from ERP systems to the SCM system. Any changes to a master record are *not* transferred back to the ERP system.

3.3.1.1 Integration of the Vendor Master Record

For source determination, SAP SCM requires vendors. The vendor master data is taken from the ERP system as a location having the location type "vendor". The leading system for master data maintenance is primarily the ERP system. Changes to the master data are transferred to SAP SCM during the next replication. As a rule, SAP ERP should be the primary system for the maintenance of vendor master data. However, for visualization and the precise graphic imaging of the supplier network on a map, geographical data such as longitude and latitude of the vendor location can be entered into the SAP SCM system.

3.3.2 CRM Middleware

In computer science, *middleware* refers to an application-neutral program that mediates between two applications. Thus, middleware is basically a distribution platform that enables data exchange between decoupled application components.

Data exchange between SAP ERP and SAP CRM is conducted via CRM Middleware. The task of this middleware is the controlled replication, synchronization and distribution of master and transaction data between the connected systems. It supports the services of initial and delta data exchange (see below).

The initial data exchange between SAP CRM and SAP ERP, also called *initial download*, includes the replication of all master data objects and system settings (customizing) from an existing ERP backend system. To enable cross-system business processes via CRM and backend systems, the following objects in particular are exchanged:

- **System settings**
  This refers to *customizing data.*
- **Master data**
  Such data includes business partner and business partner relationships, as well as plants from the ERP system and material master that are replicated as products in the CRM system.
- **Condition data**
  This data includes all information from the ERP system required by SAP CRM for the condition technique. This information consists of customizing settings and the individual condition records.

In a productive system environment, transaction data must be sent from SAP CRM to SAP ERP and vice versa in real time. The so-called *delta data exchange* denotes a continuous exchange of data between the two systems. Transaction data includes customer orders that can initially be sent from SAP ERP to SAP CRM and then, during operation, be exchanged in real time between the systems. In addition to customer orders or service documents, invoices are replicated between the systems within the context of CRM invoicing. For this, delivery data is sent from ERP Delivery to Invoicing in the CRM server. After invoicing documents are generated, this data is sent back to SAP ERP to update the status and document flow of that delivery.

3.4 Summary

Organizational structures serve to map an existing corporate structure in an SAP system and, using their individual elements, to delimit corporate divisions that need to be mapped differently due to their logistical processes. In this chapter, we have explained the primary organizational structures necessary to understand the information in the next chapter. There, we will examine the use of organizational structures within a logistics context, and detail their control and outline functions in the various processes.

Master data refers to data sets that remain unchanged over an extended period and are stored centrally in the database. This chapter has illustrated the most important master data for logistics. This data includes the business partners involved in logistics processes, such as customers and vendors, as well as the logistically relevant material master records and their conditions for distribution and procurement. The general significance of this master data and its distribution among the individual systems of SAP Business Suite were examined.

The process-specific significance and use of these and other types of master data, their relevance, and their characteristics in the logistical core processes of external procurement are the focus of the following chapter.

Chapter 4
Procurement Logistics

Procurement logistics is a segment of logistics spanning business processes from goods procurement to the transport of materials to a receiving storage location or production site. Procurement logistics connects the distribution logistics of the external supplier with the production logistics of one's own company. Its primary tasks within the framework of need-based procurement are to make available all goods and services necessary for planned operational performance processes in the correct type, quality and quantity.

Control of Business Material Flow
Procurement logistics is the first link in the logistics chain and the start of the control of business material flow. This material flow can be executed in the same way as internal procurement, through stock transfer or in-house production, or via external procurement with the aid of a vendor.

In this chapter, we will provide an introduction to the basic functions of procurement and the core processes necessary to understand procurement logistics. Thus, we limit this topic to the external procurement of goods, raw materials and supplies, and the determination of possible sources of supply. (The procurement of external services is not within the scope of this chapter.) The actual requirement for an external procurement originates either within the context of consumption-based planning or in a department that signals such a requirement.

Requirement and sales planning are discussed in detail in Chap. 5, "Production Logistics". Procurement logistics focuses on the connection between production logistics and the acquisition of determined requirements, supply source management and purchasing. The last portion of this chapter deals with the monitoring of goods receipt and payment.

J. Kappauf et al., *Logistic Core Operations with SAP*,
DOI 10.1007/978-3-642-18204-4_4, © Springer-Verlag Berlin Heidelberg 2011

4.1 The Basics of Procurement Logistics

In procurement logistics, purchasing is the first step in an optimized, process-oriented supply chain, and serves to supply a company and its production process with raw materials and supplies of the periodically recurring requirement. Such procurement objects include component supplies, semi-finished products and trading goods—but generally not finished products.

4.1.1 Business Significance

The business significance of *materials management*, as a component of logistics, comprises the entirety of materials-related tasks having to do with the supply of a company and the control of material flow, including production and delivery to the customer. According to its classical definition, its purpose is to ensure constant availability of the necessary materials—in the correct quantity and quality, at the right time, in the right place and at the right cost. The major difference between traditional and integrated materials management has already been explained in the preface of this book.

Other aspects to consider are long- and short-term security of supplies (the delivery reliability and long-term quality capability of the supplier), as well as various cost components. In addition to procurement volume, such components also include internal and external transport costs, storage and capital lockup costs, costs of the commissioning, recycling and disposal procedures, and the costs of materials management and information systems.

The following represent functions of materials management in connection with procurement and storage:

- **Traditional, external materials management**
  Traditional, external materials management primarily focuses on purchasing in terms of making supplies available within the bounds of the law, and on actual procurement logistics, consisting of the external transport, delivery and subsequent storage of the procurement objects.
- **Integrated materials management**
  The tasks of *integrated materials management* include internal production logistics such as procurement and storage, but also movement, provision, distribution and disposal. Integrated materials management thus consists of an operational supply system spanning all value-enhancement levels of a company from the vendor to the customer.

In its entirety, materials management thus comprises the most important tasks of procurement logistics—procurement, storage and provision—as well as remaining materials recycling and disposal.

Therefore, procurement logistics can be seen as the part of materials management that has to do with the physical acquisition and provision of required materials and goods. The core function of procurement in this regard can be divided into procurement planning, determination of qualitative, quantitative and time-related materials requirement, and the purchase of externally procured objects. Further tasks include supply source management, the search for potential vendors, requests for quotations, evaluation and the supervision of scheduled contract fulfillment.

Stockkeeping includes the planning of a warehouse and its furnishings, warehouse management, control of material movement and the physical execution of warehouse tasks. These include goods receipt and a quality check of goods placed into and removed from stock, as well as additional internal warehouse processes, such as inventory. Actual provision involves commissioning, transfer and internal transport, with the transport logistics goal of providing the required material to the right place at the right time as inexpensively as possible.

All tasks accruing in the following processes are mapped in the SAP ERP system:

- Planning (requirements planning) to support procurement procedures and determine the required materials
- Procurement of materials and/or services for commerce, administration, production and internal use, as well as quantitative and value-based invoice verification (interface with Financial Accounting)
- Stockkeeping and warehouse management, inventory management of internal and external materials and the analysis of materials on the balance sheet

In this book, and especially in this chapter, we would like to present the process-oriented view of procurement logistics and its mapping with components in SAP Business Suite. Thus, we have separated procurement logistics with regard to its stock and provision functions from actual procurement, and dedicated a separate chapter to both of these subareas of materials management (Volume II, Chapter 3, "Warehouse Logistics and Inventory Management").

4.1.2   Systems and Applications of External Procurement

From a business standpoint, *Purchasing*, as part of the ERP materials management component MM (Materials Management), relates to the operative activities dedicated to supplying a company with the goods and services that it requires to conduct the production process and that it does not manufacture itself. The following sections provide an overview of which SAP systems can be involved in the purchasing and procurement processes, which tasks they perform and how their integration contributes to an optimized external procurement process.

4.1.3   SAP ERP Materials Management

Within SAP ERP, there is the main component *Logistics* (LO), which is divided into further components such as Materials Management, Sales and Distribution, and Production Planning and Control. Materials Management is further divided into the subcomponents Purchasing, Inventory Management and Invoice Verification.

Thus, SAP ERP supplies comprehensive and convenient functions in the realm of procurement logistics covering the processing, optimization, monitoring and analysis of process-oriented supply chains. With a view to the procurement cycle (see Fig. 4.1), the functions in SAP ERP can be divided into the following phases:

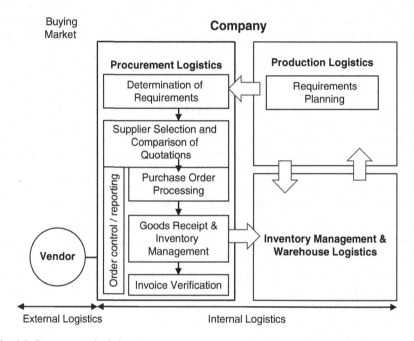

Fig. 4.1 Procurement logistics

1. **Determination of material requirements**
   The determination of material requirements sets off the procurement cycle. This can occur either automatically or manually (by a department). A requirement is defined by the quantity, point in time and location of its source. A material requirement can originate directly in a department or be the result of materials planning. In the latter case, in general business practice, a requisition note or so-called *purchase requisition* is created if the planned material is to be obtained externally. In the case of internal procurement, the SAP system generates a plan-driven order or production order to prompt the manufacture of the required amounts.

2. **Source determination**
   Purchasing supports the external procurement of materials through its selection of a vendor. This can be done under consideration of past orders or based on existing contracts. The actual determination of requirements and the integration of procurement logistics into the materials planning stage of production logistics is explained in Sect. 4.3.
3. **Vendor selection and comparison of quotations**
   The object of vendor selection and the comparison of bids is to determine the most inexpensive purchasing conditions, especially when something is being procured for the first time. The procedure generally involves creating vendor quotation requests and comparing various offers.
4. **Order processing**
   After a vendor is selected, order processing comprises all activities that turn a purchase requisition and an existing quotation into an external procurement in the form of a purchase order. As soon as an order is sent to an external vendor or another division of the company, the order represents a request to deliver the materials in accordance with stipulated conditions.
5. **Purchase order monitoring**
   Following order processing, purchase order monitoring involves the monitoring of all external procurement procedures as well as the analysis and observation of the purchase order history. Such procedures include the monitoring of goods receipts, invoice receipts and delivery costs.
6. **Goods receipt and inventory management**
   Goods receipt—the acceptance of requested materials from an external vendor or another corporate division—completes the actual procurement procedure with the testing of permissible tolerances and quality, and leads to an increase in inventory.
7. **Invoice verification**
   At the end of a logistics chain consisting of purchasing and inventory management, invoice verification checks the subjective, price- and accounting-related accuracy of the vendor invoices, and informs the responsible accountant of any deviations in quantity and price.

   In addition to these core processes of external procurement, SAP Supply Chain Management offers applications closely integrated with the ERP processes that supplement procurement processes, especially with regard to cooperation with external service providers and procurement planning.

4.1.4   SAP Supply Chain Management

Nowadays, many companies rely on a cooperative relationship with service providers, component suppliers and customers. This approach helps them to react to market demands more quickly and flexibly, to handle current innovations and shorter product life cycles, and reduce costs.

Supply chain management (SCM) includes the administration of material, information and capital flow in a network consisting of suppliers, manufacturers, distributors and customers. For effective supply chain management, the coordination and integration of this flow within a company and between companies is decisive.

SAP Supply Chain Management, as part of Business Suite, is a comprehensive solution that enables companies to efficiently plan and achieve products and services through their entire life cycle by supporting synchronized and close interaction between all departments within a logistics chain.

This logistics chain involves customers, sales and distribution, product planning and warehouse management as well as purchasing and suppliers. It places special emphasis on the processing of cross-company, cooperative planning and procurement processes.

Chapter 5, "Production Logistics", describes the fundamentals of procurement planning with SAP SCM in this regard. This section deals with the significant integration points of SCM-based procurement planning with ERP-based external procurement.

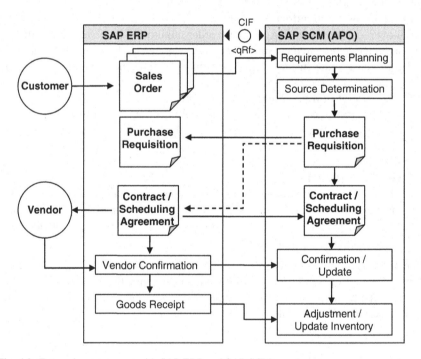

Fig. 4.2 External procurement with SAP ERP and SAP SCM

Requirements planning and the determination of the supply source are done in the SAP SCM component SAP APO (*SAP Advanced Planner & Optimizer*) (see also Fig. 2.9). Figure 4.2 shows how requirements planning and the source of supply determination can be performed by the APO system. The master data required for this procedure is typically generated in the ERP system and replicated in the APO system via CIF (see also Sect. 3.3.1 for an explanation of *APO Core Interface*). Requirements planning, controlled through SAP APO, and its integration with the ERP system are done as follows:

1. Actual requirements planning is conducted in the APO system, whereby the replicated purchasing information records from SAP ERP represent external procurement relationships in SAP APO. (Information on such purchasing info records can be found below in Sect. 4.2.3.)
2. The result of requirements planning is usually a purchase requisition that is transferred to the ERP system.
3. Once there, the purchase requisition is transformed into a purchase order that communicates with the external vendor and is transferred to SAP APO.
4. The external vendor notes the order and, depending on the stipulated type of confirmation, sends an order confirmation or shipping notification. The required quantities are then reduced in SAP APO by the confirmed amounts.
5. All goods receipts for that particular order are recorded in SAP ERP.
6. As soon as the goods receipt for the order has been recorded, the received goods are automatically transferred to SAP APO. The inventory is then increased and the open purchase order quantity is reduced.

In the course of developing traditional supply chains into supply networks, collaboration and transparency are increasingly important. For manufacturers and component suppliers, successful collaboration begins by ensuring simple information access regarding inventory for all participants, so that component suppliers always know when and what their customers require.

4.1.5   SAP Supplier Network Collaboration

SAP Supplier Network Collaboration (SAP SNC) is an SCM component and part of SAP APO. As an Internet-based software application, it enables improved collaboration with external suppliers and ensures that your supply network becomes faster, more precise and more flexible (see Fig. 4.3). SAP SNC provides inventory information quickly and seamlessly, and allows external suppliers to react independently to certain requirement and inventory situations. Therefore, SAP SNC is particularly ideal for producing companies with make-to-stock production (such as the automotive industry) and commercial enterprises.

Especially for replenishment planning outsourced to a supplier, or *Supplier Managed Inventory* (SMI), SNC provides the supplier with all information necessary for that supplier to plan the replenishment supply. For this purpose, the SNC system connected from SAP ERP is not only supplied with current inventory information, but also with the planned requirement. Based on this information, every examined period provides information about what quantity the supplier intends to deliver to the customer. For this, the supplier sets the planned procurements such that the projected inventory is always situated between a minimum and maximum inventory stored in the system. When this target inventory is calculated, the system takes into account the current warehouse stock as well as the planned additions and outward movement.

The result of this planning is procurement. In such a case, the supplier can independently create an order in SAP ERP (see Fig. 4.3) or directly create a shipping notification for the order or planned acquisition. Posting of the goods is done in much the same way as external procurement. We will explain this procedure in Sect. 4.5.

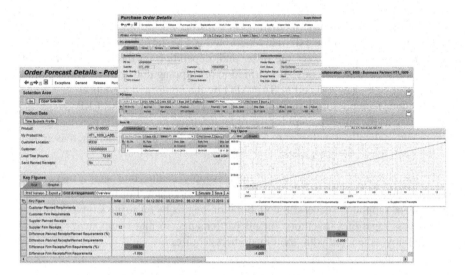

Fig. 4.3 SAP Supply Network Collaboration (SAP SNC)

Figure 4.3 shows the browser-based interface of SAP SNC. In this example, the supplier was given the current inventory and requirement situation. Based on the displayed inventory range, the supplier can create an order directly in the customer system and edit documents already generated by him or her at any time.

4.1.6   SAP Supplier Relationship Management

SAP Supplier Relationship Management (SAP SRM) offers innovative methods for the coordination of business processes with your key suppliers, and aids in increasing the efficiency of these processes. Through the system-supported optimization of your purchasing strategies, you can work together with suppliers more effectively and profit from all supplier relationships on a long-term basis.

SAP SRM supports you in examining and forecasting operative purchasing behavior and shortens procurement cycles through real-time collaboration with external suppliers. This will enable your processes to become more efficient, and allow you to reduce procurement expenses and work more closely with your suppliers.

As part of SAP Business Suite, SAP SRM represents a central component for the procurement of goods and services, and contains comprehensive reporting functions as well as electronic catalogs and their respective maintenance tools (SAP Catalog and Content Management), a vendor integration tool (Supplier Self-Service), and bid invitation and auction functions.

Although this chapter on procurement logistics primarily focuses on external procurement with SAP ERP, we would like to illustrate below the close integration of SAP SRM and SAP ERP and the differences between the functions of the two systems.

SAP ERP offers the following procurement functions:

- Operative and tactical procurement
- Contract management
- Order processing
- Invoice verification
- Procurement of external services
- Integration with transport and warehouse logistics

Thus, SAP ERP is the central system for *operative order processing*. It offers the original functions of operative purchasing as well as a seamless integration with inventory management and financing and invoicing.

SAP SRM offers the following procurement functions :

- Central source determination and management of operative contracts
- Strategic source determination with bid invitations
- Strategic source determination with live auctions
- Supplier qualification
- Vendor evaluation
- Analytical functions

SAP SRM offers the following business scenarios:

4.1.6.1   The Management of Operational Contracts

Using the *Central Contract Management* function, purchasers from various divisions of a company in different locations can profit from the conditions of globally negotiated contracts for specific product categories.

In SAP SRM, *centrally agreed contracts* can be created, which can be used as a supply source in SAP ERP as well as SAP SRM. The relevant data for such contracts is sent as a source of supply to the ERP system, and a specific contract or delivery schedule can be generated in SAP ERP. Centrally agreed contracts can not only be created and edited, but existing contracts can be renegotiated either directly with the supplier or by creating a bid invitation. In addition, a contract can be either automatically allocated as a source or listed among several other contracts as a potential supply source contract. A strategic purchaser can create a contract as soon as he or she has planned a long-term relationship with a supplier and grant individual users or user groups certain authorization levels for contracts. Centrally agreed contracts can be distributed to purchasing organizations with access rights (see Chap. 3, "Organizational Structures and Master Data"), which can then use these sources in a corresponding ERP system. (In this regard, Fig. 4.19 shows an integrated external procurement based on an SAP SRM contract.)

You can organize, structure, display and search for contracts with *hierarchies*. If you use SAP NetWeaver Business Warehouse (SAP NetWeaver BW), you can display various consolidated reports regarding contract management. For instance, you can display the total value that has been released against all contracts in a contract hierarchy.

4.1.6.2   Strategic Source Determination with Bid Invitations

With *strategic sourcing* based on bid invitations, you can obtain material pertaining to bid invitations (i.e. a *Request for Information, Request for Proposal* and/or *Request for Quotation*). In doing so, you can either work with or without integration to source determination (using the *Sourcing application*). The Sourcing application supports professional purchasers in the processing of requests and the determination of the best source of supply. Once tenders have been received from suppliers, you can create an order or contract directly from the Sourcing application or in the SAP Bidding Engine as the result of a bid invitation.

As an alternative to bid invitations, you can also employ *live auctions* for your source determination. When you use live auctions for strategic source determination, you can, for example, set rules for the submission of tenders, and bidders can submit tenders in a separate auction application in real time. This business scenario can be used with or without the support of the Sourcing application. Just as in the case of bid invitations, the bids or results of an auction can be used to generate orders or contracts.

Supplier qualification enables external suppliers to register themselves via a link on the home page of your company, where they can also allocate themselves to one or more product categories. Purchasers can then put together questionnaires that are either related to or independent of a product category, allowing them to request all general information on potential vendors. The system sends the questionnaires directly to the external supplier after they have successfully registered.

After the purchaser has accepted suppliers as potential business partners, these suppliers can be introduced to the productive purchasing system via a defined interface and noted, for instance, as a desired participant in a bidding process or as part of a supplier list. A vendor can be temporarily or permanently blocked due to inferior quality of delivered goods or performed services. In addition, the purchaser can decide whether a supplier should be authorized to independently edit his or her data und create other follow-on documents, such as invoices.

The *supplier evaluation,* as part of the SRM analytical capabilities, offers you the opportunity to evaluate your suppliers using an online survey. Surveys and questionnaires can be configured to specific requirements. You can select the criteria that you wish to assess and determine when the survey should take place. The data is transferred to SAP NetWeaver, where various reports are available that can be used to analyze results, select suitable suppliers and negotiate the best conditions.

You can create and distribute surveys using a *Supplier Survey Cockpit.* Responses coming into the system are monitored, and reminders are sent to those vendors who have not answered. The vendor rating is primarily employed to evaluate day-to-day activity based on operative documents, and thus functions to improve strategic and long-term supplier relationships.

Analytical functions allow companies to evaluate their expenses. Using these functions based on SAP NetWeaver BW, data from an entire spectrum of heterogeneous systems and from all relevant business divisions can be extracted, consolidated and made available in accordance with a company's reporting requirements. The system's flexibility enables the generation of such reports as material expenses, consolidated expense volumes, redundant vendors and possibilities of consolidating requirements.

Integrated procurement with the aid of SAP ERP *and* SAP SRM offers the following options:

- The management of catalog contents
- Self-service procurement
- Procurement with supplier integration

The *Catalog Content Management* function serves to generate internal electronic catalogs or integrate external catalogs for use in SAP ERP and SAP SRM. These catalogs are based on data in SAP NetWeaver Master Data Management (SAP NetWeaver MDM) and can be accessed via a browser in the procurement processes of SAP SRM.

Catalogs are used in external procurement to search for, compare and acquire products and services of external suppliers. Procurement using self-service catalogs aids employees in finding the right products and minimizes incorrect procurements, especially when images and descriptions are integrated.

Core Functions of SAP NetWeaver Master Data Management
SAP NetWeaver MDM supports the following:

- Functions for Catalog Content Management such as the import of catalog structures or data, the transfer of catalog items into procurement applications, and search functions
- Purchaser catalogs
- Supplier catalogs in a Web-based environment

4.1.6.3   Self-service Procurement

Self-service procurement offers employees the opportunity to create ordering procedures and enter their requirements (such as recurring requirements for office supplies or other consumable items) directly in a browser-based *purchasing portal* (see Fig. 4.4).

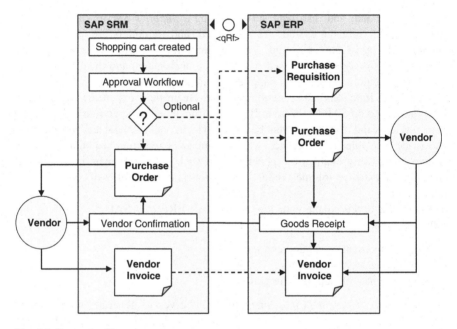

Fig. 4.4 External self-service procurement

The use of electronic catalogs speeds up the procurement process while sparing the Purchasing department from complex administrative tasks and enabling workers to dedicate their efforts to the actual supply source management.

In this self-service scenario, only the shopping cart is generated in SAP SRM. All other documents, such as orders and invoices, are managed in the ERP backend system.

The actual procurement process, depending on business requirements and the system configuration, can consist of the following steps:

1. Using a browser-based interface called the *Employee Self-Service* (ESS), an employee selects his required items from one or more electronic catalogs and places them in the virtual shopping cart.
2. The employee can then name his shopping cart (see Fig. 4.5), and verify whether or not it must be approved on the basis of operative requirements (purchasing volume, maximum cost, etc.). In such a case, an approval workflow can be automatically initiated. As soon as a shopping cart has been approved by a

manager, these requirements are sent to SAP ERP. (Figure 4.5 shows an approval workflow that can also be used for the approval of a shopping cart.)

3. These requirements in a connected ERP system can lead to a reservation, purchase requisition or directly to an order that is forwarded to an external vendor. The decision as to which system is used to generate the purchasing documents depends on the material group of the item ordered. Generally, it is possible to have the documents generated either in SAP SRM or SAP ERP.

4. With the Supplier Self-Service function, a supplier can create an order confirmation or shipping notification for that order.

5. Goods receipt is performed in SAP ERP.

6. Using SAP Supplier Self-Service, a supplier can not only generate invoices but also check the current payment status.

The actual operative procurement is conducted in SAP ERP in much the same way as in the above example. When needed, the external vendor systems can be directly connected to a company's own procurement system. Through such integration and the use of SAP Supplier Self-Service, the internal Purchasing department can work closely with external suppliers from the order process, to the shipping notification, and right through to invoicing.

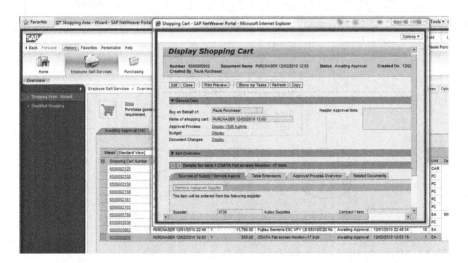

Fig. 4.5 Shopping cart in SAP SRM

4.2   Procurement Master Data

Building on Chap. 3, "Organizational Structures and Master Data", this section explains purchasing-specific functions of the material master, vendor master and purchasing info records. Figure 4.6 shows the central significance of the purchasing info record, as well as its master data relationships and the basic documents of order processing, delivery processing and invoice verification.

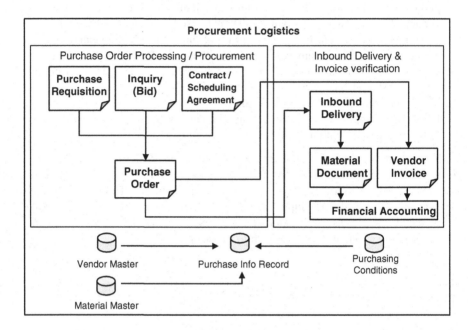

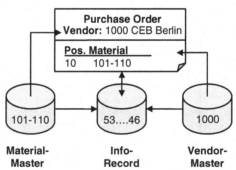

Fig. 4.6 Master data of the procurement process

4.2.1   Vendor

Information regarding the individual suppliers of a company is stored in the *vendor master records*. In addition to the vendor name and address, a vendor master record contains such details as currencies valid for a vendor, payment conditions and the names of vendor contacts.

Because a vendor is also seen from an accounting perspective as a creditor business partner of a company, the vendor master record includes accounting-related data, such as the control account of the general ledger. Maintenance of the vendor master record is thus usually performed in the Purchasing department as well as in Accounting.

Depending on business requirements, the data stored in the vendor master record may only apply to certain organizational levels. That is why the vendor master record consists of three areas, which enable a differentiated maintenance of the relevant information, divided according to company code, purchasing organization and plant. Figure 4.7 shows the views available for the maintenance of the vendor master record.

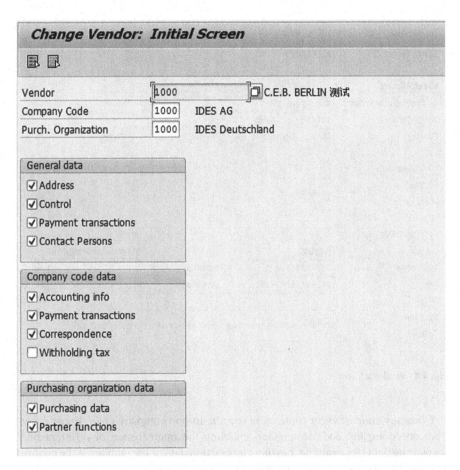

Fig. 4.7 Data views of the vendor master record

General data is data that applies equally to all company codes within a company. Such generally valid information includes the address (see Fig. 4.8), communication data (telephone, e-mail, etc.) and the language used to communicate with the supplier, as well as control data and supplier bank information.

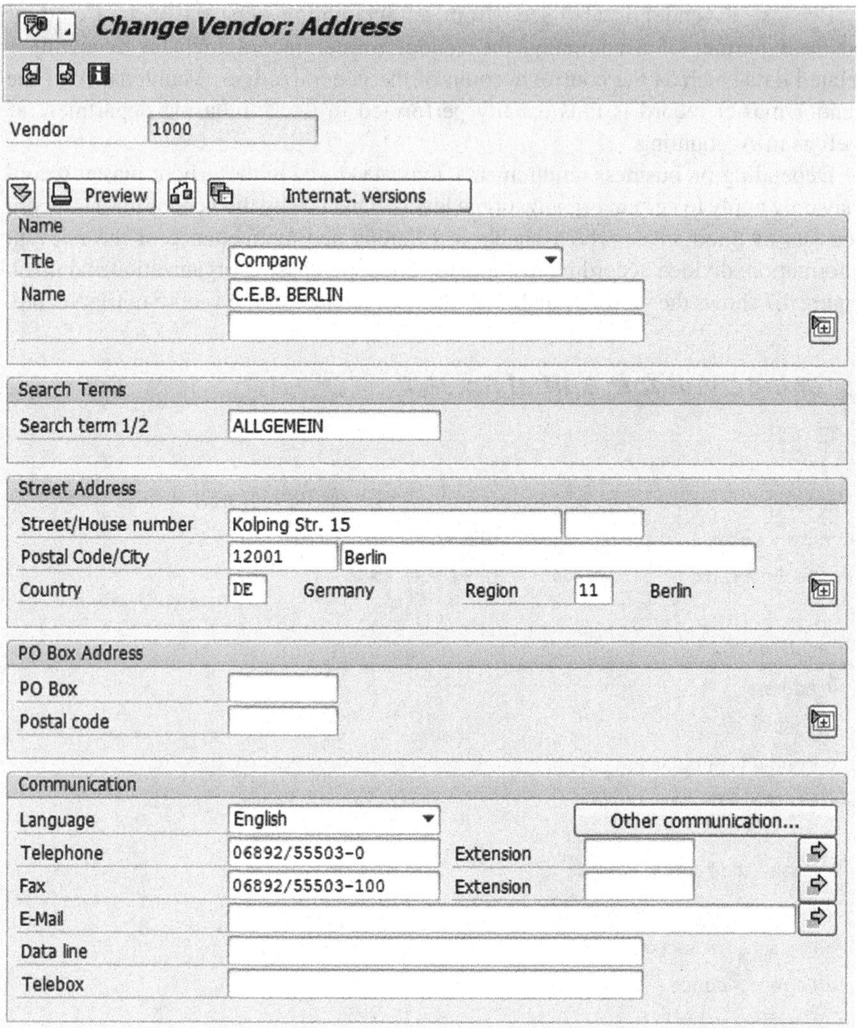

Fig. 4.8 Vendor address

Company code data, in contrast, is specific to one company code. It can differ from one company code to another, enabling the maintenance of differentiated accounting data (for example, payment transaction data or the number of the control account) in accordance with company requirements. Such data also includes information pertaining to account management and the control account.

When a vendor is created in the system, a distinct number is issued for the creditor. The way in which a number is issued, i.e. whether an external number is assigned by an accounting clerk or an internal number is issued by the system, depends on the so-called *account group* of the creditor. The assigned creditor number is also the sub-ledger number in Accounting. In the sub-ledger, the sum of payables per vendor is updated.

The data involved here represents accounting payment obligations of the company, especially stemming from external procurement procedures, and as such, a *control account* must also be maintained. The control account is a company-code-dependent master record in the vendor master record and corresponds to a general ledger account. It comprises the obligations of a firm toward several vendors in G/L accounting.

Company-code-dependent data (see Fig. 4.9) contains account information required for internal control, data on payment transactions, i.e. information regarding stipulated payment conditions for payment of vendor invoices and regarding correspondence with vendors, as well as optional information on withholding tax.

The actual *purchasing data* depends on the purchasing organization, and represents the next level of data in the organization-dependent vendor master record. The data needed for purchasing in your company, which can be organized according to purchasing organization, generally includes the respective supplier contact person and general delivery and order conditions, which can be used as recommended values for purchasing info records and outline agreements.

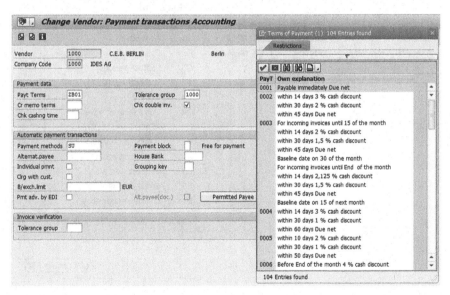

Fig. 4.9 Vendor accounting data

The order currency and payment conditions are entered in the purchasing data of the vendor master record. In addition, important control data is also stored in the vendor master record. Various codes decide on the use of further purchasing functions, such as the automatic generation of orders from purchase requisitions or automatic goods receipt settlement.

In addition to purchasing data, the vendor master record can also contain *partner functions* (see Fig. 4.10). Partner functions are business partners that take on certain functions at an external vendor location in the procurement process. Such functions include, for instance, an alternative invoicing party that needs to be identified when booking an incoming vendor invoice with regard to an order for that vendor.

Generally, business partners can take on various roles regarding a company, depending on the business requirement. During a procurement procedure, the vendor master record determines the ordering address of a company, then the actual goods supplier, subsequently the invoicing party, and finally the recipient of the payment. Required for the use of partner functions is a corresponding master record for the respective partner and the maintenance of the relationships, that is, the partner function, in the respective vendor master record. Partner functions are assumed in purchasing documents as recommended values.

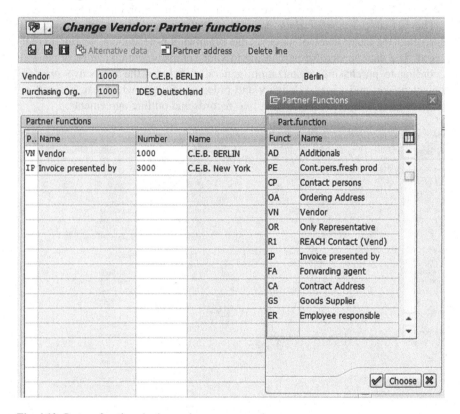

Fig. 4.10 Partner functions in the vendor master record

Divergent Partner Functions in a Vendor Master Record
Figure 4.10 shows the vendor master record of Vendor 1000. The supplier, the firm C.E.B. in Berlin, has the partner function VN (vendor). The actual recipient of payment, defined by partner function IP (invoicing party), is Supplier 3000, the parent company of the supplier, C.E.B. in New York.

In addition to the data that applies to a purchasing organization, other purchasing data or partner functions can be maintained for a specific plant or supplier sub-range that deviates from purchasing organization data, such as payment conditions or Incoterms. Data that deviates from purchasing organization data is called *alternative data*.

A vendor master record is a prerequisite to generating a purchasing document (such as a request for quotation, purchase order or outline agreement). In the event that a vendor master record has not yet been created for a business partner, a suspense account or so-called *CpD vendor master record* can be used. A CpD master record (*Conto pro Diverse*) is utilized for several vendors when no master record is to be created for them, especially for suppliers to whom one-time orders are placed. In contrast to the vendor master records already explained, it is thus not possible to save supplier-specific data in the master record for CpD vendors.

In the previous chapter, in the discussion of the company code view of a debitor, we saw that the account group selected for a vendor master record is controlled by its number assignment. When a CpD vendor master record is created, a special account group for CpD vendors is assigned. This account group hides the vendor-related fields. Therefore, this data must be entered manually when a purchasing document, such as a purchase order, is created.

4.2.2   Material

For external procurement, the focus of this chapter, the use of a material master record is not always necessary. Although a material master record is not required for materials whose consumption is assigned to cost centers or customer orders, the master record is necessary for external procurement of stock material—especially due to the required quantity and value update.

The material master record includes the descriptions and control data of all articles and parts that a company procures, manufactures or stores. The material master record is the central source for the retrieval of material-related information. The integration of all material data in a single master record eliminates the problem of redundant data and enables joint use of the stored data by the Purchasing department as well as other divisions of a company.

The basic features and views of the material master record were presented in Chap. 3, "Organizational Structures and Master Data". In this section, we will provide a brief explanation of the purchasing-specific information in the material master record that can be maintained either on the client or plant level.

Data on the *client level* applies equally to every company, plant and warehouse within a business. *Plant data* is relevant to the individual plant locations or departments within a company. *Purchasing-specific data* is generally maintained on the plant level.

This purchasing-specific, plant-related data primarily includes purchasing data—that is, material master data provided by the Purchasing department. Such information includes the *purchasing group* responsible for the procurement of material, the quantity of the permissible overdelivery and underdelivery tolerances and the actual *purchase order unit*, in the event that it deviates from the client-wide valid *base unit of measure* (see Fig. 4.11).

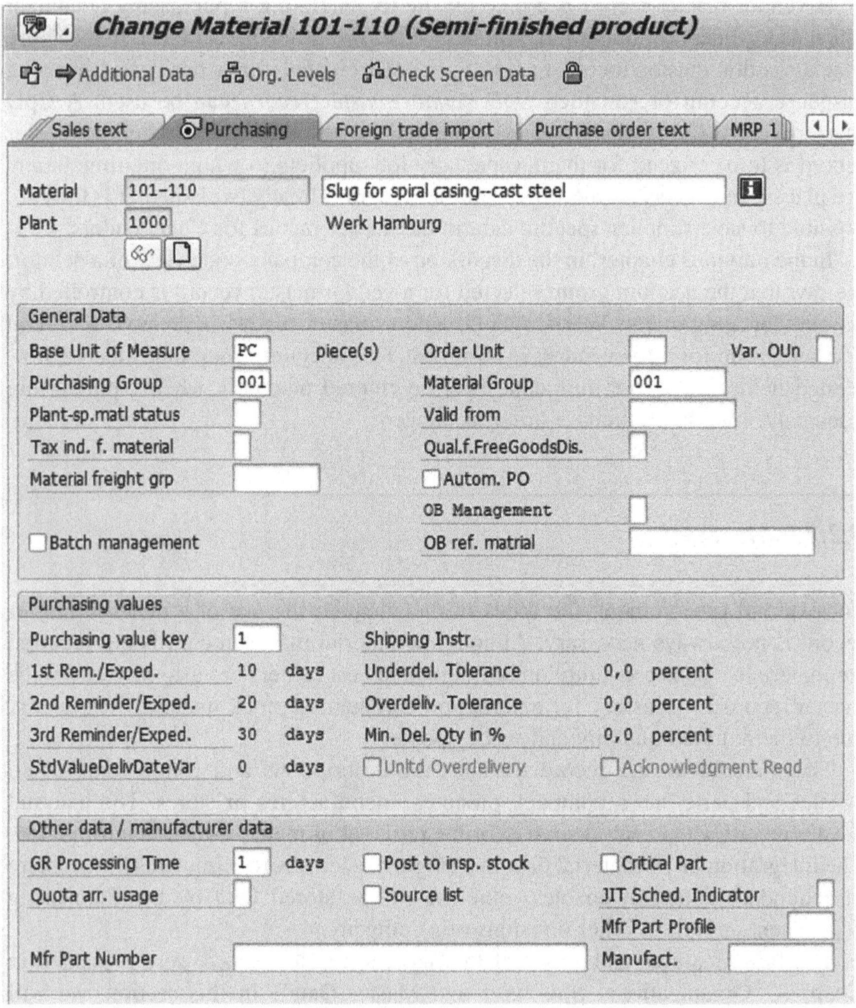

Fig. 4.11 Purchasing view of the material master

In addition to the material description, the material master can also be used to enter language-specific *purchase order texts*—under the tab "Purchase order text"—which provide a more detailed description of a material or include important information for the supplier. One text can be entered per language. This text is automatically assumed in the various purchasing documents, depending on the language of correspondence of the supplier, and can be edited manually (see Fig. 4.12).

The *planning views* contain information on material requirements planning, safety stock and reorder levels, and planned delivery times of a material (see Fig. 4.13).

In this regard, the scheduling types maintained in the planning views contain information about the procurement type of a material.

The *procurement type* indicates how the procurement is to be executed—whether as in-house or external production or whether both types are possible. If, in accordance with this setting, both types of procurement are possible and you do not enter a quota arrangement, the system assumes in-house production. The planning run thus initially generates planned orders that you can then convert into production orders or purchase requisitions. You will find more information on the integration with requirements planning in the following sections. Quota arrangement as an instrument for purchasing optimization is described in the section "Supply source determination" (Sect. 4.6.1).

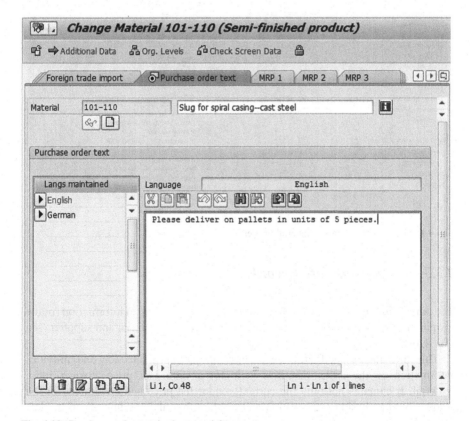

Fig. 4.12 Purchase order text in the material master

The *special procurement type* indicates how in-house or external procurement is to be done: through stock transfer, external production, consignment, etc.

Under the tab Foreign trade: import, the material master record contains plant-related foreign trade data and import codes, as well as the country of origin and results of supplier confirmations, customs preferences for the respective zones, and data pertaining to legal controls and the numbers of any existing clearance certificates. More detailed information on foreign trade procedures in SAP Business Suite can be found in Volume II, Chapter 4, "Trade Formalities – Governance, Risk, Compliance".

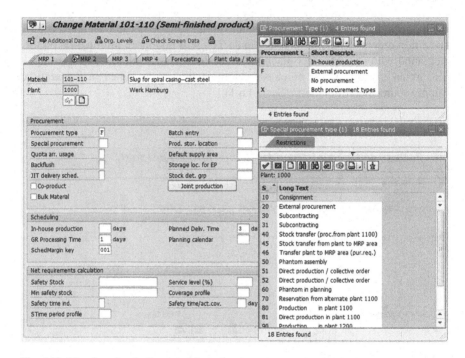

Fig. 4.13 Planning view in the material master

4.2.3   Purchasing Info Records

The purchasing info record (or info record for short) serves as an information source for purchasing, and establishes the relationship between material and supplier (see Fig. 4.14).

This info record facilitates the process of quotation selection and provides information on the order quantities valid for a vendor or price changes for a material. It contains not only the current vendor list price and conditions for the responsible purchasing organization and/or plant, but also the number of the last order. Based on this information, Purchasing can ascertain at any time what materials have been previously offered or supplied by a certain vendor and under which conditions.

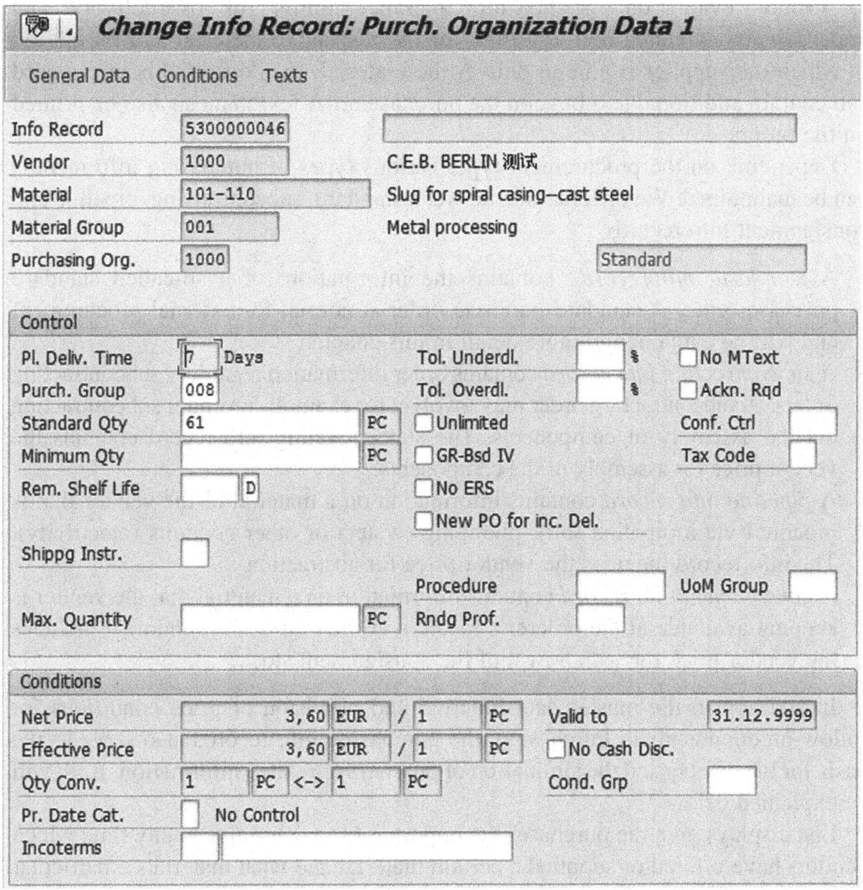

Change Info Record: Purch. Organization Data 1

General Data   Conditions   Texts

Info Record	5300000046	
Vendor	1000	C.E.B. BERLIN 测试
Material	101–110	Slug for spiral casing–cast steel
Material Group	001	Metal processing
Purchasing Org.	1000	Standard

Control

Pl. Deliv. Time	7 Days	Tol. Underdl.	%	☐ No MText
Purch. Group	008	Tol. Overdl.	%	☐ Ackn. Rqd
Standard Qty	61 PC	☐ Unlimited		Conf. Ctrl
Minimum Qty	PC	☐ GR-Bsd IV		Tax Code
Rem. Shelf Life	D	☐ No ERS		
		☐ New PO for inc. Del.		
Shippg Instr.				
		Procedure		UoM Group
Max. Quantity	PC	Rndg Prof.		

Conditions

Net Price	3,60 EUR / 1 PC	Valid to		31.12.9999
Effective Price	3,60 EUR / 1 PC	☐ No Cash Disc.		
Qty Conv.	1 PC <–> 1 PC	Cond. Grp		
Pr. Date Cat.	No Control			
Incoterms				

Fig. 4.14 Purchasing organization data in the purchasing info record

Planned Delivery Time and Conditions
Figure 4.14 shows Purchasing Info Record 5300000046, which describes the relationship between Material 101–110 and Vendor 1000. According to this master record, the planned delivery time amounts to 7 days and the material can be acquired for €3.60 per unit.

The current and future quotation conditions (discounts, fixed costs, etc.) are entered into the info record and can be used in a procurement purchase order. The vendor info record contains the net quotation and order data—condition data, used as recommended values in purchase orders—as well as data pertaining to vendor evaluation, permissible tolerance limits for overdelivery and underdelivery, planned delivery time for the respective material and the period in which the supplier is able to deliver the material. In addition, this info record can contain additional text beyond the purchase order text that can also be printed on the orders.

Depending on the procurement type, various types of purchasing info records can be maintained. We differentiate between standard, subcontracting, pipeline and consignment info records:

- A *standard info record* contains the information for a so-called standard purchase order. A standard purchase order is normal for external procurement and will be explained in more detail in this chapter.
- A *subcontracting info record* contains order information regarding subcontracting orders. A subcontracting order may involve, for example, a vendor subcontracting out the assembly of components. The subcontracting info record contains the vendor price for assembly of the components.
- A *pipeline info record* contains information on a material of the vendor that is procured via a pipeline (oil), plumbing (water) or other conduits (electricity). The info record includes the vendor price for abstraction.
- A *consignment info record* contains information on a material that the vendor is keeping available at the orderer's location at the cost of the vendor. It includes the vendor price for withdrawal of the consignment stock.

In addition to the master data function and provision of price conditions for follow-on documents in Purchasing, the purchasing info records also serve as the basis for list display and the simulation of purchasing-related information. Both will be explained below.

List displays give the purchaser the opportunity to determine at any time which vendors have offered or supplied a certain material and what materials a particular vendor can deliver under which conditions. In addition, purchasing info records provide insight into the order and quotation price history, and enable a net price simulation for material to be obtained externally.

Using the price simulation function, Purchasing can compare the prices of various vendors for a material or material group, as well as the prices of materials of one particular vendor, and have the system calculate the net price of a material offered by several vendors depending on a certain simulated quantity.

A further evaluation based on purchasing info records is the order price history, with which Purchasing can immediately find out about a vendor's price changes for a particular material (see Fig. 4.17). In the order price history, the various prices a

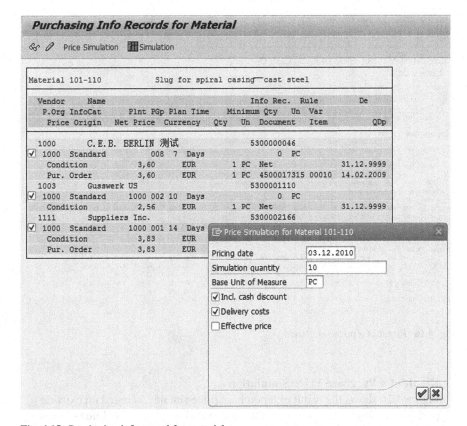

Fig. 4.15 Purchasing info record for material

Purchasing Information for Material
Figure 4.15 shows the purchasing information for a material. Material 101–110 can be acquired from three different vendors. The existing purchasing info records allow the purchaser to perform a price simulation. For this, he or she selects the purchasing info records to be compared, enters the quantity required and determines whether the system should take into account any scaled prices as well as discounts and delivery costs. The result of the simulation is the net price, the lowest of which represents the least expensive procurement option.

vendor charges for a material are recorded by generating a price history record for every order item that is based on an info record. This price history documents any deviation in price for that item.

Price Simulation for Info Records

🔲 🔍 🔽 🔽 🔳 ⅗ ⟳ 🖥 ⟳ 🖥 ⟳ 🔲 🔲 🔲 🔲 🔲 | 🖩 &Condition Record &Material &Vendor &Info Record &Outline Agreement

Calculation of ;; Net Value Inclusive Cash Disc. and
Delivery Costs

Material 101-110
Quantity 10 PC
Date   03.12.2010

P	Vendor	POrg	Plant	Info Record	Text	Net Value	Crcy	CTyp	Name	Rate	Crcy	per	UoM	Cond.value	Material	Matl Group	Q	Fix
Calculated Value	**1:**	**34,92 EUR**																
	1000	1000		5300000046	Standard	34,92	EUR	PB00	Gross Price	3,60	EUR	1	PC	36,00	101-110			
	1000	1000		5300000046	Standard	34,92	EUR		Net incl. disc.	3,60	EUR	1	PC	36,00	101-110			
	1000	1000		5300000046	Standard	34,92	EUR		Net incl. tax	3,60	EUR	1	PC	36,00	101-110			
	1000	1000		5300000046	Standard	34,92	EUR	SKTO	Cash Discount	3,00-	%	0		1,08-	101-110			
	1000	1000		5300000046	Standard	34,92	EUR		Actual Price	3,49	EUR	1	PC	34,92	101-110			
	1000	1000		5300000046	Standard	34,92	EUR		🔲Border crossing			0			101-110			
Calculated Value	**2:**	**24,83 EUR**																
⊙⊙🔲	1003	1000	1000	5300001110	Standard	24,83	EUR	PB00	Gross Price	2,56	EUR	1	PC	25,60	101-110			
⊙⊙🔲	1003	1000	1000	5300001110	Standard	24,83	EUR		Net incl. disc.	2,56	EUR	1	PC	25,60	101-110			
⊙⊙🔲	1003	1000	1000	5300001110	Standard	24,83	EUR		Net incl. tax	2,56	EUR	1	PC	25,60	101-110			
⊙⊙🔲	1003	1000	1000	5300001110	Standard	24,83	EUR	SKTO	Cash Discount	3,00-	%	0		0,77-	101-110			
⊙⊙🔲	1003	1000	1000	5300001110	Standard	24,83	EUR		Actual Price	2,48	EUR	1	PC	24,83	101-110			
⊙⊙🔲	1003	1000	1000	5300001110	Standard	24,83	EUR		🔲Border crossing			0			101-110			
Calculated Value	**3:**	**37,15 EUR**																
	1111	1000	1000	5300002166	Standard	37,15	EUR	PB00	Gross Price	3,83	EUR	1	PC	38,30	101-110			
	1111	1000	1000	5300002166	Standard	37,15	EUR		Net incl. disc.	3,83	EUR	1	PC	38,30	101-110			
	1111	1000	1000	5300002166	Standard	37,15	EUR		Net incl. tax	3,83	EUR	1	PC	38,30	101-110			
	1111	1000	1000	5300002166	Standard	37,15	EUR	SKTO	Cash Discount	3,00-	%	0		1,15-	101-110			
	1111	1000	1000	5300002166	Standard	37,15	EUR		Actual Price	3,72	EUR	1	PC	37,15	101-110			
	1111	1000	1000	5300002166	Standard	37,15	EUR		🔲Border crossing			0			101-110			

Fig. 4.16 Result of a price simulation

Result of a Purchase Price Simulation
Figure 4.16 shows the result of a purchase price simulation based on existing
purchasing info records for Material 101–110. Order conditions were
stipulated with three vendors. Vendor 1003 offers the best conditions for
the procurement quantity of 10 units under consideration of possible scaled
prices, at an actual price of €2.48.

Purchase Order Price History

🔍 &Info Record

Info Rec.   Vendor   Material   P.Org Plnt InfoCat
5300000046 1000   101-110   1000   1000 Standard

Date	Net Price	Curr.		Qty	Un	Order No.	Item	Variance
☐16.05.1997	8,18	DEM	/	1	PC	4500004477	00010	
	4,18	EUR	/	1	PC			
☐13.01.2009	5,00	EUR	/	1	PC	4500017307	00010	19,6
☐23.01.2009	3,60	EUR	/	1	PC	4500017311	00010	28,0-

Fig. 4.17 Order price history for a purchasing info record

4.3 Determination of Requirements and External Procurement

The actual requirements planning for material to be externally procured is part of operational planning and is explained in more detail in Chap. 5, "Production Logistics". Requirements planning can be done in SAP ERP as well as SAP SCM (APO component). The following section provides an overview of the process integration of the determination of requirements and its result, the requisition note and purchase requisition.

4.3.1 Integration with Requirements Planning

The central task of requirements planning is guaranteeing material availability by procuring the necessary amount internally and in a timely manner for distribution. Using various planning methods and procedures, the system determines shortage situations and automatically generates the respective procurement proposals for Purchasing or production.

With materials requirement planning, we can generally differentiate between *consumption-based* and *deterministic* planning. The planning type depends on the corresponding parameters in the material master. The decisive factors here are the plant-related codes of *planning characteristic* and *planned lot size*. They enable a material to be planned in different plants with different planning types.

Information on Sales Planning
A detailed description of sales planning can be found in Chap. 5, "Production Logistics".

We will briefly discuss the integration of procurement with consumption-based planning, which, in contrast to deterministic MRP, is solely oriented on the internal consumption of a material (see Fig. 4.18).

External requirements, such as customer orders, planned independent requirements and reservations, do not affect planning in such cases. Consumption-based planning is thus mainly practiced in companies that do not have their own production or for the planning of category B or C materials. Deterministic MRP is primarily suited to the planning of category A materials, i.e. finished products.

Consumption-based planning is based on past demand values and thus has is not related to any production plans. The actual requirement—depending on the selected planning procedure—is determined with the aid of a forecast or statistical procedures. The latter are triggered by dropping below a set reorder level, the so-called *reorder point*, or by forecast requirements calculated from past consumption.

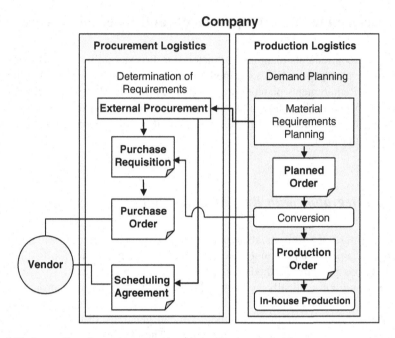

Fig. 4.18 Integration of procurement with consumption-based planning

Three procedures are available for consumption-based planning: reorder point planning, forecast-based planning and time-phased materials planning.

- **Reorder point planning**
  For reorder point planning, procurement is always triggered when the sum of the current plant stock and planned inflow falls short of the *reorder point*. The reorder point is chosen such that it covers the expected average material requirements during the replenishment lead time. In order to compensate for any excess consumption or delivery delays during the actual replenishment lead time, a safety stock can be defined. The safety stock is thus a part of the reorder point, and can be manually defined during manual reorder point planning, depending on the on-time delivery performance of a supplier, any forecast errors that need to be considered, or based on a targeted service level. In automatic reorder point planning, the safety stock is automatically determined by an integrated forecast program. The forecast values for future requirements depend on material consumption up to that point. Since the forecast program can be executed at regular intervals, the reorder point and safety stock match the respective consumption and delivery situation. This contributes to inventory reduction.
- **Forecast-based planning**
  Forecast-based planning is also oriented on the material consumption of previous periods. Like automatic reorder point planning, this type of planning also uses forecast values for future requirement generated by an integrated program.

However, unlike reorder point planning, these forecast values serve as the requirement quantities for the planning run, and, as forecast requirements, have a direct effect on material requirements planning. The main advantage is that the automatic forecast calculation, which is performed at regular intervals, uses data from the past to forecast future materials requirement, which is adapted to current consumption patterns.

- **Time-phased planning**
  If a vendor always supplies a material in a certain rhythm, such as on a particular day of the week, it is a good idea to perform the planning of that material using the same rhythm and deferred by the delivery period of the vendor. This planning time phase is stored in the material master.

Actual planning is normally done on the *plant level*. This means the system calculates the plant stock, with the exception of individual customer stock, as the sum of stock from the various storage locations. However, it may be necessary to either refrain from making certain storage location stock available for plant planning or to plan it separately.

These organizational units, which are to be planned independently, are called *MRP areas*. There are two basic types of MRP areas:

- The *plant MRP area* includes the plant that is to undergo MRP planning and all allocated storage locations. If individual storage locations are planned as independent MRP areas, the plant MRP area is reduced by those storage locations in order to prevent redundant planning.
- For the *MRP area for storage locations*, individual storage locations are allocated to one MRP area. For requirements planning, these storage locations are then separated from the rest of the plant and planned collectively.

A *procurement proposal* for replenishment of the calculated shortage is automatically generated by the system during the planning run, and determines how procurement of that material is to be done. We differentiate between in-house production and external procurement:

- **In-house production**
  For *in-house production*, or internal procurement, the system generates *planned orders* to plan the quantities to be produced. After the planning procedure is complete, the planned orders are converted to corresponding *production orders* and forwarded to the Production department.
- **External procurement**
  For *external procurement*, either a *planned order* or a *purchase requisition* can directly be generated. Once planning is complete, the planned order is converted to a *purchase order* and forwarded to Purchasing. The advantage of first generating a planned order for external procurement is that the planner obtains an additional check of the procurement proposals. Purchasing cannot order until the planned order has been converted to a purchase requisition. If a purchase requisition is directly generated, Purchasing assumes the responsibility for material availability and warehouse stock.

An exception is *stock transfer processing*, in which goods are procured and delivered within a company. The plant that is to receive the goods orders them internally from another plant, which delivers the goods. Stock transfer processing using a stock transfer order is done when two plants are far away from one another because in this case, transport of the materials to be transferred must be considered in the planning.

If a scheduling agreement exists for a material and if there is a planning-related entry in the source list, schedule lines can also be directly generated by requirements planning. In this regard, the schedule line can be done either by SAP ERP or SAP APO, depending on the system employed.

External procurement with SAP APO based on APO scheduling agreements is executed in the following steps:

1. The APO scheduling agreement in SAP ERP links to SAP APO as the planning system, where it generates a so-called external procurement relationship. This external procurement relationship is allocated to a transport relationship and thus represents the relationship of the external supplier to the procuring plant.
2. Actual requirements planning with source determination is done in SAP APO based on customer orders and reservations from the ERP system (see also Fig. 4.2). Such planning results in allocations in a scheduling agreement.
3. The delivery schedules and their transfer to an external supplier can either be created interactively or automatically via a background job.
4. The supplier has the opportunity to display and confirm the delivery schedules over the Internet, via a so-called *Supplier Workplace* (SWP).
5. The individual schedules are transferred to the ERP system and saved as allocations in the scheduling agreement.
6. SAP ERP then generates a delivery, which, when transmitted to SAP APO, reduces the allocations and delivery schedules by the notified quantity.
7. Goods receipt is done in SAP ERP and increases the stock in both systems while reducing the allocations and schedules in SAP APO.

An APO scheduling agreement differs from that of ERP in that the complete delivery processing continues to be executed in SAP ERP. In SAP APO, the ERP scheduling agreement only serves as an external procurement relationship that can be allocated to a purchase requisition in the planning run.

All further steps, such as converting the purchase requisitions into scheduling agreement allocations and the generation of a delivery schedule, are done in the ERP system. We will explain the scheduling agreement as an outline agreement in an optimized purchasing process in Sect. 4.6. In this context, Fig. 4.50 will show you the basic procedure of scheduling agreement control with and without release order documentation.

One distinctive feature for which all of the systems mentioned in this chapter are used is integrated external procurement with SAP SCM and SAP SRM (see Fig. 4.19). In this scenario, requirements planning is performed with SAP APO. This planning results in procurement requisitions for which a supply source is determined. If the determined supply source is an operational contract that was generated in SAP SRM, the procurement requisition or sales order is transferred to the SRM system.

Possible procurement elements and thus the result of planning include:

- Planned orders for materials that are externally procured or produced in-house.
- Procurement requisitions for externally procured materials.
- Scheduling agreement allocations for externally procured materials for which an entry has already been made in the source list and a scheduling agreement already exists.

Scheduling agreement allocations for outline agreements leads to the release of the required quantities from the supplier. Planned orders are either converted to production orders or purchase requisitions. The following describes the purchase requisition as a trigger for procurement.

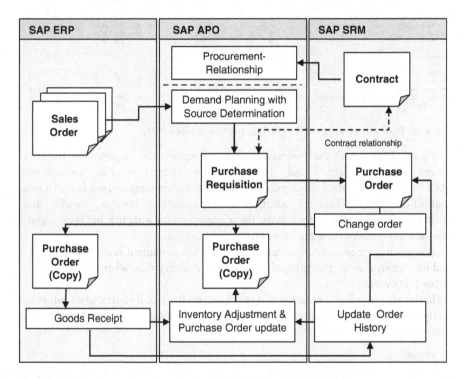

Fig. 4.19 External procurement process with SAP APO and SAP SRM

4.3.2 Purchase Requisition

A purchase requisition represents a request to the Purchasing department to procure a material or service in a specified quantity and at a specific point in time. Purchase requisitions (abbreviated as *PurRqs*) are either directly or indirectly generated and are always internal documents. They are not passed on to external suppliers in that form and are solely for internal use.

In relation to the procurement process, a purchase requisition can be seen as its trigger. Every purchase requisition can contain a *requirement tracking number* for definitive identification of the requirement and for tracking of the resulting purchasing transaction.

Purchasing Transactions per Requirement Tracking Number

Purch. Doc. Cat.	Item	Material	Short Text	Type	Doc. Date	I	A	D	Plant	SLoc	Quantity	OUn	PGr	Matl Group	SPlt	Number	Σ Number
👁																	••• 13
Req. Tracking Number BEVERAGE																	•• 8
Purchasing Document 10014060																	• 2
Purchase Requisition	10	R100003	Lemonade 1.5 l bottle	NB	03.12.2010				R320	0001	10	PC	R30	R1112			1
Purchase Requisition	20	R100022	Bottle 1.5	NB	03.12.2010				R320	0001	100	PC	R30	R1221			1
Purchasing Document 10014061																	• 3
Purchase Requisition	10	R100003	Lemonade 1.5 l bottle	NB	03.12.2010				R120	0001	5	PC	R30	R1112			1
Purchase Requisition	20	R100022	Bottle 1.5	NB	03.12.2010				R120	0001	19	PC	R30	R1221			1
Purchase Requisition	30	R100003	Lemonade 1.5 l bottle	NB	03.12.2010				R120	0001	55	PC	R30	R1112			1
Purchasing Document 4500017470																	• 3
Purchase Order	10	R100003	Lemonade 1.5 l bottle	NB	03.12.2010				R320	0001	0,833	CRT	R30	R1112			1
Purchase Order	11	R100022	Bottle 1.5	NB	03.12.2010				R320	0001	9,996	PC	R30	R1221			1
Purchase Order	12	R100023	Crate 12 bottles	NB	03.12.2010				R320	0001	0,833	PC	R30	R1221			1
Req. Tracking Number FOOD																	•• 5
Purchasing Document 10014062																	• 3
Purchase Requisition	10	R100026	Meyer's cream of mushroom soup	NB	03.12.2010				R315	0001	450	PC	R30	R1114			1
Purchase Requisition	20	R100027	Meyer's goulash soup	NB	03.12.2010				R320	0001	11	PC	R30	R1114			1
Purchase Requisition	30	R100028	Meyer's soup display	NB	03.12.2010				R120	0001	1	PC	R31	R1114			1
Purchasing Document 10014063																	• 2
Purchase Requisition	10	R100028	Meyer's soup display	NB	03.12.2010				R320	0001	4	PC	R31	R1114			1
Purchase Requisition	20	R100004	'Sophia L' pizza, 3-pack	NB	03.12.2010				R320	0001	10	PC	R30	R1113			1

Fig. 4.20 Purchasing transaction requirement tracking numbers

Figure 4.20 shows the purchasing transactions for the requirement tracking numbers for "Beverage" and "Food" and the related purchase requisitions 10014060 and 10014061. Both purchase requisitions are summarized in collective requisition 4500017470. In addition to the material number, vendor and corresponding purchase order history, the requirement tracking number is an important selection criterion for the monitoring of a purchase.

If a purchase requisition is directly generated, the document is entered manually and the material to be procured, along with the delivery date when the material is needed, is entered.

In the case of indirect creation of a purchase requisition, it is generated automatically by another SAP component. Automatic generation can be triggered by the following procedures and processes:

- **Planning**
  In the ERP system, the consumption-based planning recommends the materials to be procured based on past consumption and existing stock. The result of planning externally procured materials is a purchase requisition. In addition, for SCM-based requirements planning, a purchase requisition can be generated through the conversion of planned orders in the ERP system. Integration in the requirements planning has already been examined in the previous section.
- **Distribution logistics**
  If a customer order contains one or more items that are not in stock and must be obtained externally, a purchase requisition can be automatically generated in

Purchasing. A detailed description of such individual orders is given in Chap. 6, "Distribution Logistics".

- **Supplier Relationship Management**
  In much the same way as self-service external procurement can be done via SAP SRM, a purchase requisition can also be generated in EAP ERP after a shopping cart has been created in SAP SRM and approved.

In addition to the integration options cited above, purchase requisitions can be automatically generated through so-called *networks* in the ERP component PS (Project System) as well as via *maintenance orders* of the component EAM (Enterprise Asset Management, Maintenance and Service Management). *Production orders*, which are generated in production planning and control, can also generate purchase requisitions if they contain services, usually external processing or non-stock components.

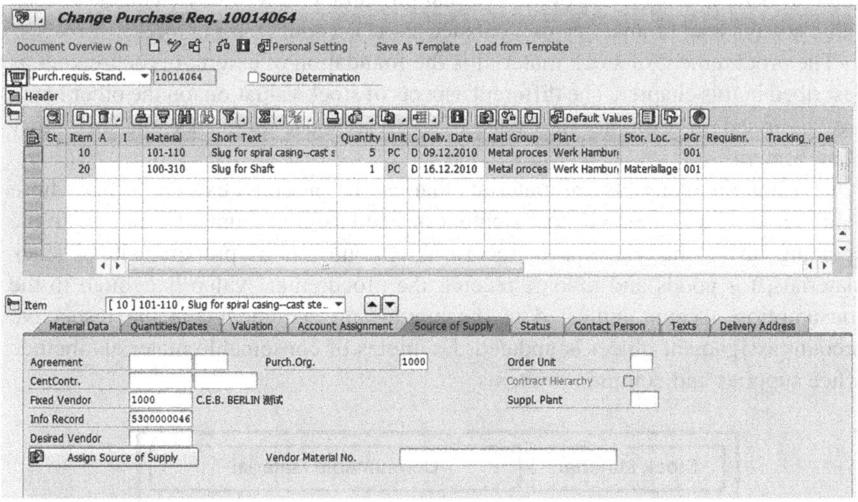

Fig. 4.21 Purchase requisition

Purchase Requisition
Figure 4.21 shows a purchase requisition having the number 10014064. This purchase requisition contains two items that are materials to be procured externally. Item 10 (5 units of Material 101–110) was generated with reference to the existing purchasing info record 5300000046, and is to be procured externally from Vendor 1000, the company C.E.B. in Berlin.

The purchase requisition contains the items to be procured along with the respective procurement types. Possible procurement types include "normal" procurement, that is, external procurement, as well as procurement through subcontracting or external service, vendor consignment and stock transfer. As with stock transfer, vendor consignment is examined as special stock in Volume II, Chapter 3, "Warehouse Logistics and Inventory Management". In addition to the procurement type, the item also indicates the quantity and delivery date for the material to be supplied as well as the scope of required services.

Depending on whether the materials to be procured externally are stock or consumable materials, certain information is either compulsory or optional on the item level. A purchase requisition can be created for material with or without a material master record. If a material is to be obtained for a cost center, an account assignment category is issued for that item.

Purchase requisition items for the external procurement of stock material require the indication of a material number. This material number forms the basis of quantity, value and consumption updates in the material master record. The connection to a material master record also enables the record to be updated with regard to the moving average price of the external procurement, whereby changes in stock value always lead to an entry in a balance sheet account (see Fig. 4.22).

The procurement of stock material is the foundation of external procurement as described in this chapter. The different aspects of stock valuation, on the other hand, are discussed in Volume II, Chapter 3, "Warehouse Logistics and Inventory Management".

Consumable materials are materials that can be procured externally but whose value settlement is done via cost element or fixed asset accounts. In contrast to the updating of the material master record as is done for the procurement of stock materials, for goods and invoice receipt, the procurement value is debited to the consumption account indicated in the purchasing document and the respective account assignment object is updated. Examples of consumable materials include office supplies and computer systems.

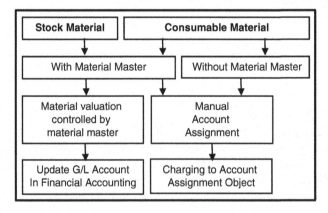

Fig. 4.22 Account assignment of stock and consumable material

The procurement of consumable materials does not necessarily require entering a material number. Here it is important to differentiate between:

- Consumable material without a material master record.
- Consumable material with a material master record for which neither quantitative nor value-based materials management is conducted.
- Consumable material with a material master record for which a quantitative but not value-based materials management is conducted; this includes the procurement of stock material for internal consumption.

For the procurement of consumable materials without a material master record, master data attributes, such as description and order units, cannot be determined from a material master and must be manually entered in the document. In contrast to the procurement of stock material, acquisition of consumable materials requires assignment to an account assignment object (such as a cost center, projects, plants, or directly to a customer or production order), since material consumption is generally booked as an expense. There is no value, quantity or consumption-based update.

Purchase requisitions may be subject to an approval procedure. Depending on the business necessity, if certain conditions are fulfilled, a purchase requisition must be approved before further processing is allowed. The release procedure will be explained in Sect. 4.6.4.

4.4   Order Processing

Order processing, the actual purchasing process, is the heart of procurement logistics. Figure 4.23 shows the integration of order processing with procurement logistics and the downstream processes of delivery and invoice verification.

For external procurement, typical order processing starts with a request for quotation issued to a vendor and the actual quotation. The subsequent purchase order, which is forwarded to the vendor, is usually confirmed by that vendor, and the system generates a delivery. The delivery is the basis for goods receipt and the generation of a material document. Invoice Verification can then check the invoice received by the vendor and finally pay the bill.

4.4.1   Request for Quotation and Vendor Quotation

A request for quotation is a request to a vendor to submit a quotation on the supply of materials or the performance of services.

The request for quotation is an optional, manual step that can either be done manually or in the context of an existing purchase requisition or outline agreement. Using a request for quotation, you can manage and compare the quotations of vendors.

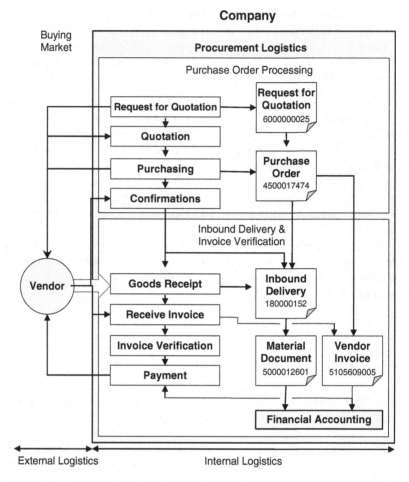

Bestellabwicklung_Übersicht

Fig. 4.23 Overview of order processing and delivery

Figure 4.24 displays a request for quotation to a vendor that has been created in connection with a purchase requisition. The corresponding purchase requisition data, such as the delivery data, the materials to be procured and the quantity, are automatically assumed in the request for quotation.

The request for quotation to a vendor consists of the RFQ header and RFQ items, citing the materials to be externally procured, as well as the corresponding delivery dates. Unlike the subsequent purchasing documents and preceding purchase requisition, the request for quotation cannot indicate account assignment.

In Purchasing, the request for quotation and quotation are one and the same document. The RFQ is created, then subsequently printed out and sent to the external vendor.

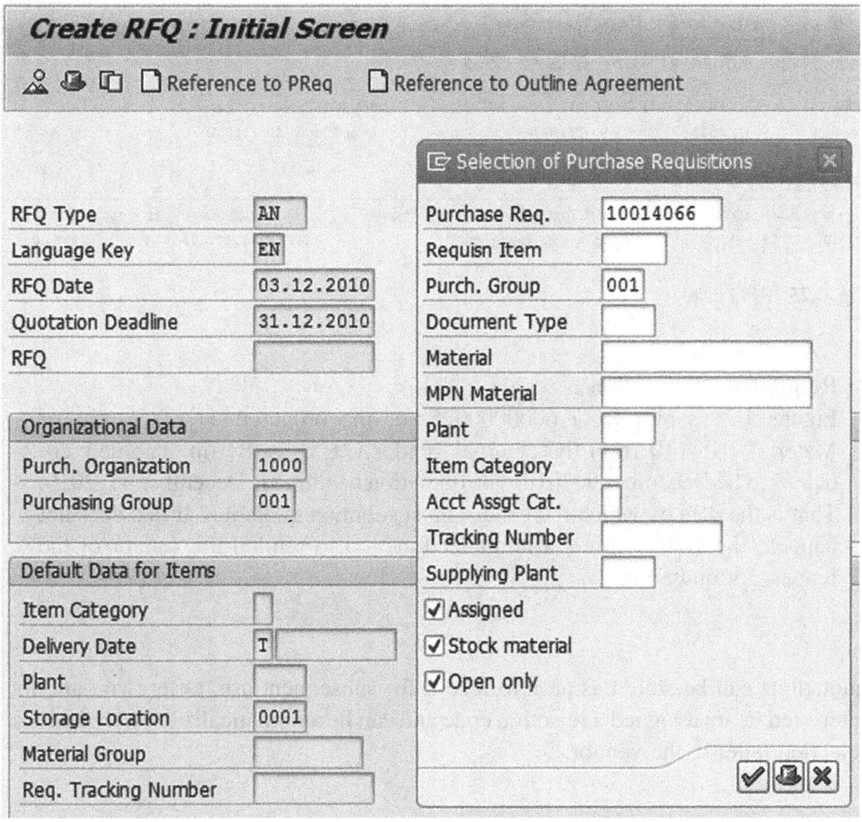

Create RFQ : Initial Screen

👤 🖨 ▢ ▢ Reference to PReq   ▢ Reference to Outline Agreement

		▷ Selection of Purchase Requisitions
RFQ Type	AN	Purchase Req.	10014066
Language Key	EN	Requisn Item	
RFQ Date	03.12.2010	Purch. Group	001
Quotation Deadline	31.12.2010	Document Type	
RFQ		Material	
		MPN Material	
Organizational Data		Plant	
Purch. Organization	1000	Item Category	
Purchasing Group	001	Acct Assgt Cat.	
		Tracking Number	
Default Data for Items		Supplying Plant	
Item Category		☑ Assigned	
Delivery Date	T	☑ Stock material	
Plant		☑ Open only	
Storage Location	0001		
Material Group			☑ 🖨 ✖
Req. Tracking Number			

Fig. 4.24 Creating a request for quotation (RFQ)

The prices and conditions based on the request for quotation to an external vendor are updated in the original RFQ—in this way, the request becomes the quotation, that is, the offer from the vendor to supply the requested materials at certain conditions. The vendor's quotation is legally binding and relevant for subsequent procurement and procurement conditions. Alternatively, the vendor can also issue a rejection for the received request for quotation. The RFQ is an optional, not a compulsory, document. Orders can generally also be created without a previous request for quotation to a vendor.

If the desired quantity is to be supplied according to a specified plan, schedule lines can be added for the individual items to record precise information on delivery date and time for individual subsets. Several related RFQs can be bundled under a *collective number*. The collective number represents an internal reference with which several RFQs and quotations can be individually selected for evaluation purposes.

If a request for quotation is sent to more than one vendor, the system can determine the cheapest offer and automatically send a rejection to the more expensive vendors. In addition, prices and delivery conditions for individual

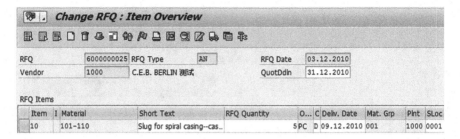

Fig. 4.25 RFQ display

Request for Quotation
Figure 4.25 shows RFQ 6000000025 for the procurement of 5 units of Material 101–110 from the external vendor C.E.B. in Berlin. The delivery date is to be December 9, 2010. The quotation deadline is December 31, 2010. That is the date by which the vendor must submit a quotation. If this does not happen, the system offers a reminder function to remind the vendor of the request for quotation.

---

quotations can be stored as an info record for subsequent use. Offers you are not interested in are assigned a rejection code and can be automatically answered with a rejection letter to the vendor.

4.4.2 Purchase Order

Within the context of external procurement, we have already explained above the determination of requirements, its integration with requirements planning, the creation of a purchase requisition and source supply determination using RFQ and quotation processing.

The purchase order represents a request to the external vendor to supply a certain quantity of a specific material at a specified time under stipulated conditions.

In addition to a purchase order for external procurement, there are other types of orders. This includes stock transfer orders, for which the material to be procured can be obtained within a company or from another organizational unit—another plant or storage location.

Information on Stock Transfer Orders
Stock transfer orders are explained in Volume II, Chapter 3, "Warehouse Logistics and Inventory Management", in the section "Goods Movement".

A purchase order can be created manually or—in order to reduce data-entry work and entry errors—directly in relation to preceding documents. The various types of order creation below are illustrated in Fig. 4.26.

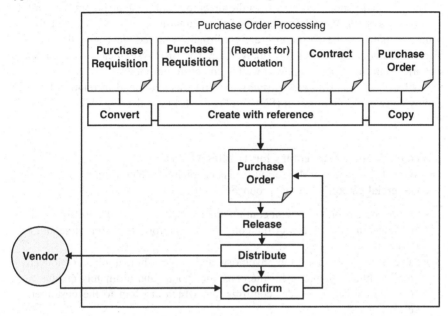

Fig. 4.26 Purchase order creation

4.4.2.1   Creating Purchase Orders

- **Manual creation**
  Manual creation of a purchase order, for which the vendor for external procurement is already known and need not be automatically determined by the system.
- **Conversion of purchase requisitions**
  Purchase orders can be directly generated through the *conversion* of purchase requisitions. Conversion can either be done manually or automatically, in which case the vendor cited in the supply source of the purchase requisition is assumed. If the purchase requisition is not yet allocated to a source of supply, a purchase order can still be created; the external vendor must then be manually entered in the order.
- **Creation of orders with reference**
  A purchase order can be created with reference to preceding documents. If reference is made to a purchase requisition, RFQ or contract, item data and, if available, header data from the preceding document will be used in the order. Creating a purchase order with reference to an existing contract is called a *release order*. Reference to a preceding document leads to an update of the purchase order item with the corresponding document and item number. Using this document flow, it is possible to check whether a reference was made, and if so, which document item was referenced in the generation of a purchase order item.

A special form of order generation is done in *Vendor Managed Inventory* (VMI), in which a supplier takes on the planning tasks for his or her articles in a customer's company.

The external vendor has access to the inventory and replenishment data in the customer's system. Actual generation of the purchase order in such a scenario is done automatically based on an order confirmation issued electronically via EDI *(Electronic Data Interchange)* by the vendor. The ERP system can achieve a VMI scenario from the supplier view as well as the customer view.

VMI or *Supplier Managed Inventory* (SMI) was explained in relation to SAP Supply Network Collaboration (SNC) in Sect. 4.1.4 (see also Fig. 4.3).

Vendor Managed Inventory for Trading Goods
A typical case where VMI is employed is the planning of consumer goods in a commercial enterprise by the producer of said goods.

- From the customer view, a continual electronic transfer of inventory and replenishment data flows from commercial enterprise to external vendors via EDI.
- The vendor receives the inventory and replenishment data, conducts replenishment planning for the consumer goods and ultimately creates a customer order and order confirmation, which are sent to the customer electronically.
- From the customer side, the order confirmation can be used to automatically generate a purchase order. The order number of this purchase order is transferred to the external vendor. Delivery of the ordered consumer goods is done in relation to this order number.

Limit orders are another special form of orders. They offer considerable advantages in the procurement of consumable materials. Normally, when consumable materials are procured, a purchase order is generated for every procurement procedure, which will subsequently serve as the basis for invoice verification. For limit orders, a purchase order has an item-related value limit, and a certain validity period is set.

A limit order represents what is known as a *blanket purchase order*, and enables consumable materials or services to be procured when individual order processing would not be wise for economic reasons. In contrast to the process overview in Fig. 4.23, this type of external procurement directly allows an incoming invoice with reference to a limit order without having to previously enter a goods receipt. The major advantages of this type of order lie in the reduced process costs, since several consumable materials and/or services can be procured over a length of time with only a single order item, without needing to generate purchasing documents for it or book goods receipts.

A purchase order consists of a document header and the order items. The document is displayed as a *single-frame transaction*. This display shows all significant data areas, header and item data on a single entry screen (see Fig. 4.27).

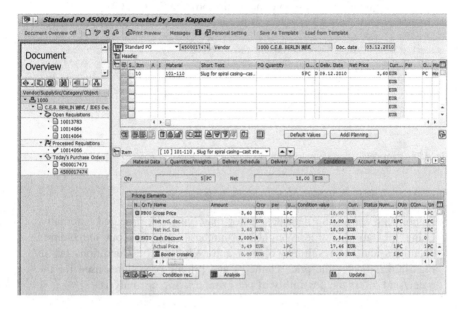

Fig. 4.27 Order document overview

Document Overview
Figure 4.27 shows an order addressed to Vendor 1000 and the document overview for the selected vendor with the completed purchase requisitions. In this example, the displayed Order 4500017474 is generated with a reference to Purchase Requisition 10014066. Three further purchase requisitions for the same vendor have not yet been processed. The item detail shows the calculation and purchasing conditions that apply for external procurement of that material from Vendor 1000.

The left margin, in the document overview screen, shows the various sales documents that pertain to an order or vendor or are required for daily work. Depending on the desired document type and time period, the document display can be limited to purchase orders, purchase requisitions, inquiries or delivery plans.

The *header* of the purchase order contains data relevant to the entire document. Such data includes payment conditions, Incoterms, organizational data and the actual source of supply, the vendor. In addition, the order header includes communication data and purchase order texts that can be forwarded to the vendor.

The *item overview* contains the most important material data, such as the material number, order quantity and the plant to which the material is to be

delivered. The item data includes the item type as well as information regarding
account assignment for which any necessary value-based update, especially for the
procurement of consumable materials, is to be undertaken (see also the explanation
on the procurement of consumable materials above in Sect. 4.3.2).

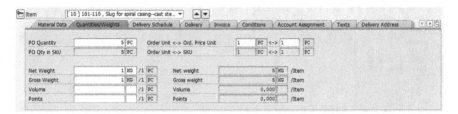

Fig. 4.28 Item details for a purchase order

Item Details for a Purchase Order
Figure 4.28 shows the item details for material to be procured externally.
Using the tab Quantities/Weights, a purchaser can find out the ordered
quantities and the resulting volumes and weights. The Conditions tab in
Fig. 4.28 indicates the price elements and pricing procedures used for the
order, and offers the possibility of performing a pricing analysis.

In this regard, the *item type* controls whether the indication or recording of a
material number, account assignment, goods and invoice receipt is possible or
necessary for the respective order item. Generally, SAP ERP recognizes the following item types in an order that requires different order processing:

- **Normal**
  This item type is used for stock or consumable material that is to be procured
  externally.
- **Limit**
  Limit items contain a value limit for the procurement of consumable materials or
  services.
- **Consignment**
  For vendor consignment goods, the vendor provides the company with materials
  that belong to the vendor until they are withdrawn. Storage of an ordered
  consignment material does not yet lead to an obligation toward the external
  supplier. The obligation only exists once the material has been withdrawn from
  the consignment warehouse.
- **Subcontracting**
  For subcontracting, a finished product is ordered from an external supplier,
  whereby the components required to produce the finished product are entered
  as order items.
- **Stock transfer order**
  This refers to an order item that is to be procured through a stock transfer.

- **Third-party order**
  A procurement process in which the ordered material is directly delivered from a vendor to a third party (such as a customer).

Stock transfer orders, consignment, third-party order processing and subcontracting are discussed in more detail in Volume II, Chapter 3. The following examples relate to "standard items" for external procurement of stock material.

The *item detail data* includes additional information regarding the respective item order (see Fig. 4.28). Such information includes further material data, such as material-specific conditions, tax codes and weights, account assignment information and tolerances for over- and underdelivery.

Figure 4.27 showed the pricing analysis for Condition PB00, the gross purchase price of Material 101–110. The purchase price of €3.60 was determined using a condition record. The condition record—in this case a material info record—comes from the purchase requisition allocated to the order, taken from the purchasing info record displayed in Fig. 4.29.

Pricing in a purchase order, the determination of the procurement price, is generally done with the condition technique mentioned in Chap. 3, "Organizational Structures and Master Data", based on so-called condition records. The condition technique, a core function in every SAP system, offers a flexible, individually controllable functionality with which to map and use price elements that occur in business practice. This not only enables price elements to be determined but also allows the identification and execution of messages, tasks and expenses based on certain criteria.

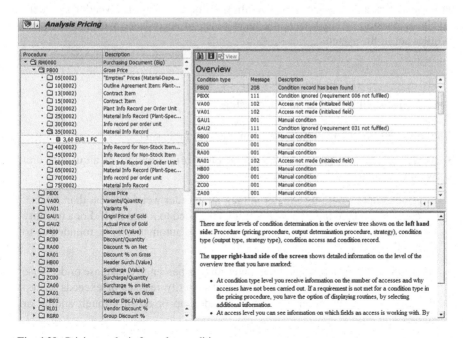

Fig. 4.29 Pricing analysis for order conditions

Information on Condition Technique
Actual condition determination in the pricing procedure is done identically in all SAP applications. It will be explained in detail in Chap. 6, "Distribution Logistics".

The special feature of pricing in Purchasing is based on the fact that purchase orders are generally created with reference to a preceding document, and the purchasing conditions in such a document can be incorporated into the pricing process. The purchase order searches for conditions in info records as well as conditions saved in outline agreements.

For automatic pricing, especially when time-based conditions are involved, the system initially bases its determination of the time-based conditions on the document date of the purchase order. Alternatively, pricing can be done based on the price control saved in the vendor master or according to the pricing date type maintained in the purchasing info record (see Fig. 4.14). These system parameters enable time-based pricing and thus the determination of purchasing conditions that are valid on the day of the purchase order, the delivery date or the date of goods receipt.

Order quantity optimization is used in ordering and contract processing to round off order quantities. Adapting order quantities can serve to optimally utilize transport capacities and n exploit the conditions stipulated with the vendor as extensively as possible. Actual rounding takes place in accordance with the rounding profiles set in the system.

4.4.2.2   Rounding Profiles in Purchasing

The following rounding profiles are available in Purchasing:

- *Static rounding profiles* can be used to round off a multiple of a quantity unit without changing the quantity unit.

  – An example would be a material that, at an order weight of less than 50 lbs., could be ordered in full pounds without the need to round it off. For a quantity of more than 50 lbs., the order quantity would automatically be rounded up to a full 100 lbs.

- *Addition and subtraction rounding* determines a percentage increase or decrease for certain threshold values. If an ordered quantity reaches or exceeds such a threshold value, the order amount is rounded up or down through adding or subtracting units.

- An example would be screws for the construction of a bookcase. Ten screws are required per bookcase. Since screws often get lost while the bookcase is being put together, the system can automatically order 15% more screws. An order of two bookcases would increase the number of screws to 23, as the order amount of 20 units is rounded up accordingly.

- A *dynamic rounding profile* can round an initial quantity unit up or down multiple times, thereby altering the quantity unit. The dynamic rounding profile is used in procurement logistics especially when a complete logistical quantity, such as an entire pallet, is to be rounded up or down.

  - An example would be a material that is delivered in one cardboard box at an order amount of more than 10 units. 100 boxes would make up a pallet. If more than 155 units of this material are ordered, the corresponding amount would be one pallet, five boxes and 5 units of the material. In this example, the order quantity unit changes from "unit" to "box" and finally to "pallet". Using the respective rounding rules in this case, the order quantity can be rounded such that it is only delivered in full pallets or boxes.

- Rounding can also be done based on a *limit value check*. The system uses this type of check after the cited rounding methods have been performed to determine the rounded order amount based on the minimum and maximum quantities in the purchasing info record (see Fig. 4.14).

  - An example would be an ordered amount of 23 units. Due to dynamic rounding, the order is increased to 30 units. The limit value check, however, indicates that the minimum quantity that can be ordered is 50 units.

As in the case of purchase requisitions, purchase orders can also be subject to an internal approval procedure. Depending on the business necessity, an order may have to be approved once certain conditions have been fulfilled before it is forwarded to a vendor. We will explain the order release procedure in this chapter in Sect. 4.6.4.

Figure 4.30 shows the possible message types for a purchase order. The messages conform to a message profile and message conditions that are specifically assigned to a certain message type.

Purchasing documents generated in SAP ERP are transmitted to an external vendor. Messages pertaining to reminders, especially a reminder to issue an order confirmation, are created via a special report for purchasing documents.

Order information is transmitted using a variety of media, such as printouts, e-mail, faxes or EDI. For this, the system generates a message for every request for quotation, purchase order, contact and respective purchasing document. This message is placed into a *spool*, which is a type of "message queue" that contains all messages not yet transmitted to the vendor. Release of these messages from the spool is either done immediately, that is, directly upon saving the purchasing document, or delayed via a so-called background job. Based on this transmitted information, the vendor can then issue the order confirmation.

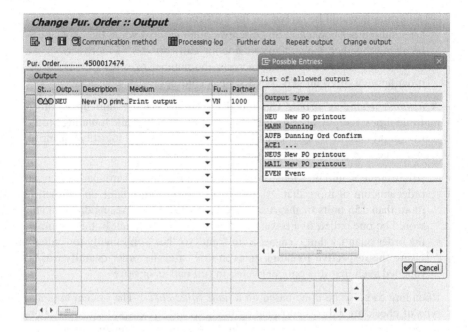

Fig. 4.30 Message output for a purchase order

4.4.3   Confirmations

Confirmations are messages from a vendor regarding the estimated delivery date and the delivered quantity of the ordered materials. Confirmations can either be created manually or automatically, based on information received electronically.

The advantage of vendor confirmations is that material requirements planning can be executed with precise information regarding delivery dates and quantities, thus enabling more exact planning, especially in the phase between the order date and desired delivery date. Confirmations also enable a more precise monitoring of a vendor's delivery reliability.

Depending on how the information is entered, we differentiate between confirmations that are directly generated in purchasing documents and those that are documents themselves. Chronologically, from the time an order is transmitted, the following confirmations can or must be exchanged in external procurement in accordance with the *confirmation control key* issued for an order (see Fig. 4.31):

1. The purchase order is transmitted to the external vendor. Transmission takes place via e-mail, fax, EDI or as a traditional printout.
2. The vendor confirms the receipt of the order and, depending on the stipulated arrangements, sends an order confirmation. The receipt of the order confirmation is directly entered in the purchasing document.

3. After the vendor has loaded the goods, a loading confirmation can then be issued.
4. The shipping notification generates its own document—the inbound delivery. The term *shipping notification* refers solely to the message itself. The shipping notification, that is, that confirmation type, then contains the name "Inbound delivery", as soon as the shipping notification has been booked in the ERP system. It is possible to work exclusively with order confirmations and create the confirmation in the item detail of the purchase order.
5. After physical receipt of the goods, the ordering party books the goods receipt.
6. The confirmations are updated in the *purchase order history* (see Fig. 4.31).

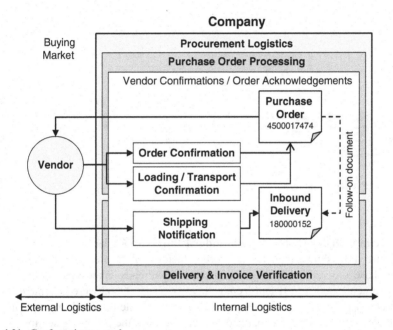

Fig. 4.31 Confirmation control

Whether or not a confirmation is to be expected for an order item and what kind of confirmation is required is determined by the confirmation control key. To make sure that an order has been received by a vendor, the ordering party can, for instance, request an order confirmation. In this case, the order confirmation has no specific confirmation control key and can be created at any time for a purchase order.

If a reminder is issued for an expected confirmation that has not been received, it is also saved as a so-called *acknowledgment requirement* per order item.

Acknowledgments from a vendor are either created manually or automatically via EDI. Depending on the confirmation type, either a separate document or a corresponding update in the purchase order is generated. Preliminary goods receipts and the shipping notification generate a material document and inbound delivery.

In the *purchase order history*, all procedures performed for an order item are documented. These procedures mark the document flow for an order item. These include goods receipts and invoice receipts, as well as important information on delivered quantities, document numbers and items, and the posting date on which the document was generated.

After the purchase order has been generated, transmitted to the vendor and confirmed by the vendor, the ERP system generates an inbound delivery.

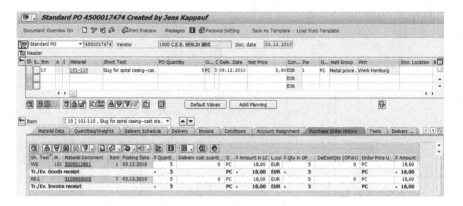

Fig. 4.32 Purchase order history

Purchase Order History of a Standard Purchase Order
Figure 4.32 shows the purchase order history of the items in a standard purchase order, for which goods receipt has already taken place. Material Document 5000012601 and Vendor Invoice 5105609005 were generated on December 3, 2010, with reference to this order item. Goods receipt and invoice verification are explained in more detail in the following section.

4.5   Delivery and Invoice Verification

Goods receipt and delivery are significant components of the logistics chain. After notification and confirmation from the vendor of the expected delivery date and quantities, delivery usually takes place and the ordered goods are stored. The end of the logistics chain for external procurement is marked by the incoming vendor invoice, which is checked in the logistics invoice verification process for factual, price-related and mathematical accuracy.

Integration with inventory management and warehouse goods receipt is described in Volume II, Chapter 3, "Warehouse Logistics and Inventory Management". The following examines goods receipt and invoice verification at the end of external procurement of stock material.

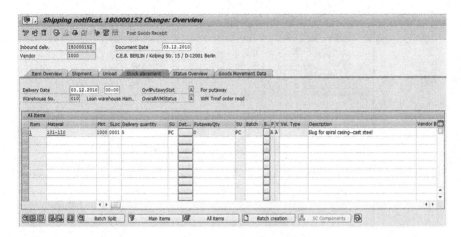

Fig. 4.33 Inbound delivery of an order

Figure 4.33 shows Inbound Delivery 180000152 for the order in the previous examples (see Fig. 4.27). This inbound delivery of 5 units of Material 101–110 takes place in Plant 1000 for Storage Location 0001.

4.5.1   Goods Receipt

Goods receipt can generally be booked with reference to a purchase order or to an inbound delivery. Using the respective preceding document, the system determines all significant information required for goods receipt:

- What? (What material was procured?)
- When? (What was the delivery date?)
- How much? (In what quantity?)
- From where? (From which vendor or, in the case of stock transfer, what delivering plant?)
- To where? (To which destination, which stock?)

In Purchasing, the purchase order is not only the central document for the external procurement of goods and services, but also an important tool for planning, inventory management and invoice verification.

In order to ensure that the delivered items are indeed what was ordered, the physical acceptance of the goods can be recorded in the goods receipt with reference to an existing purchase order (see Fig. 4.34).

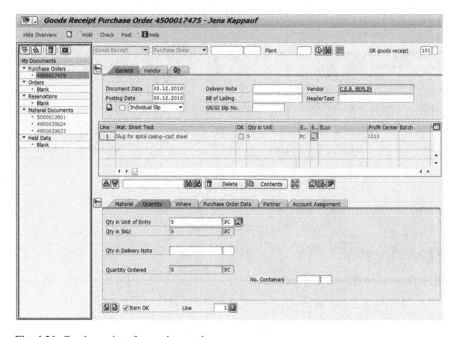

Fig. 4.34 Goods receipt of a purchase order

The system automatically recommends the data from the purchase order and, for instance, adopts material and quantity data for goods receipt.

For goods receipt, the delivered quantity is compared to the undelivered order quantity to determine the open order quantity. If an ordered quantity has been completely delivered, the system sets a *"delivery completed" indicator* in the goods receipt for that particular order item. The "delivery completed" indicator shows whether an order is considered complete and no further goods receipt is expected. The open order quantity is then set to zero.

If goods receipt and open orders do not match, it could be due to an overdelivery or underdelivery. An underdelivery corresponds to a kind of partial delivery for which, depending on business requirements, tolerances can be stored in the system. Overdeliveries are generally not permissible but can also be enabled—within set tolerances.

When a goods receipt is posted for a purchase order, the open order quantity of the order item is updated and a *purchase order history record* is automatically generated. The purchase order history record contains material movement data significant to the order, including the delivered quantity, material document number and the posting date of goods movement.

Alternatively, goods receipt can be done with reference to an inbound delivery. In this case, the shipping notification from the external vendor generates an inbound delivery as a follow-on document to the purchase order.

Goods receipts can also be entered without a reference to a purchase order or inbound delivery. Such *other goods receipts* can generate an order directly from an existing purchasing info record, depending on system settings and business requirement.

A goods receipt posting has extensive effects in the system. Firstly, a material document is generated. This document, a log for the receipt of goods, serves as proof of goods movement and leads to an updating of stock. The actual material document consists of a header and at least one document item. The header contains the posting date and the name of the author. The document items include a purchase order reference, if applicable, and the material and posted quantity, as well as information on the plant and storage location to which the stock was posted.

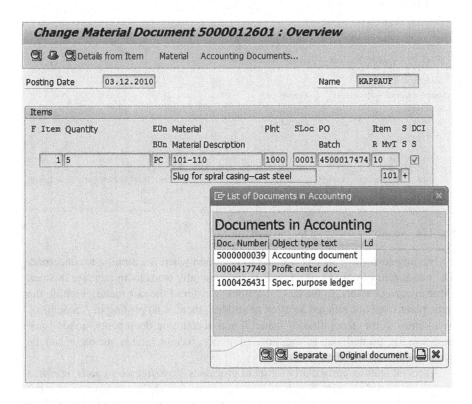

Fig. 4.35 Material document for goods receipt

Material Document for Goods Receipt
Figure 4.35 shows a material document for Purchase Order 4500017474, in
which 5 units of Material 101–110 were posted in Storage Location 0001 of
Plant 1000. In addition to the actual, physical increase in stock in this storage
location, accounting documents were simultaneously generated whose
accounting document 5000000039 follows the value-based stock change in
accounting (see Fig. 4.36).

Posting a goods receipt in a material document automatically generates accounting
lines in the determined accounts and subsequently (especially for the procurement
of consumable material) an update of the respective account assignment object.
Material and accounting documents are independent documents with their own
document numbers. Generation of both documents occurs simultaneously, whereby
the accounting document records the accounting effect of goods movement, especially
its valuation (see Fig. 4.36).

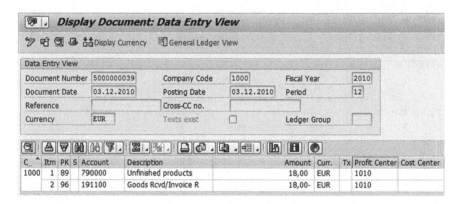

Fig. 4.36 Accounting document for goods receipt

The movement of goods is relevant to valuation when Accounting is concerned.
The goods receipt posting of a raw material generally leads to an increase in stock
value in current assets. In the case of a stock transfer of the raw material within the
same plant from one storage location to another, there is no posting in Accounting.
In addition to the generation of material and accounting documents, goods issue
slips can be printed out at the time of goods receipt, and a message can be
automatically sent to the purchaser.

A special feature, particularly for retail processes, is *preliminary goods receipt*, a
precursor to goods receipt. For this type of confirmation, the goods have arrived but
no goods receipt has yet been posted in the system. Actual goods receipt takes place
on a time delay in a second step.

This type of goods receipt is thus also known as *two-step goods receipt*, in which
the individual quantities are recorded according to the delivery note but not yet

posted. The advantage of the two-step procedure is the accuracy of the data entry, as well as the collection of all items that have been delivered by a particular vendor to a company, in order to perform tasks such as printing sales labels.

4.5.2 Vendor Returns

A return refers to the return of goods to an external or internal supplier. In the case of external procurement, a *vendor return* refers to the return of goods to an external vendor, and is generally done with a reference document. The reference document in Purchasing is the purchase order.

The result of a return in this context is a goods receipt correction and a credit against the vendor, which is taken into consideration for the subsequent invoice verification.

Returns processing in Purchasing is wholly possible on the procurement side, by identifying the order item to be returned as a return order item in the purchase order and sending the item back right at the point of goods receipt. If returns processing requires shipping documents or freight lists, the return can also be processed through dispatch handling. Dispatch handling is discussed in Chap. 6, "Distribution Logistics".

4.5.3 Invoice Verification and Handling of Payments

From a materials management standpoint, logistics invoice verification is the last accounting step of the procurement process, and involves checking the vendor invoice for factual, price-related and mathematical accuracy. Invoice verification represents the link between the ERP component Materials Management (MM) and Financing (FI).

When an invoice is posted, the ERP system sends information to Materials Management (for example, quantities), Accounting (price differences) and Cost Accounting (cost center), and updates relevant data (such as material prices). In addition, invoices blocked from payment, for instance in the case of deviations from the order document, are processed through invoice verification. Invoice verification, as a central component of purchasing, accesses all procurement logistics master data (see Fig. 4.37).

After successful processing, invoices are released to Financing for payment. SAP ERP offers various options for processing invoices:

- You can enter invoices with a reference to a purchase order. The invoice data is recorded and compared with the purchase order data. If there are deviations, for instance with regard to quantity or price, the system can assist you in determining tolerance values. After invoice verification, the invoice amount is cleared for payment.

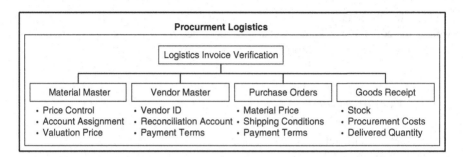

Fig. 4.37 Master data integration of invoice verification

- When the purchase order or goods receipt number is entered, invoices are automatically generated by the system and posted to the G/L and material accounts.
- Within the invoice verification process, you can also enter invoices without reference to a purchase order and post them directly to the corresponding account.

The incoming vendor invoice (see also Fig. 4.23) is recorded in the system as an invoice document. This invoice document consists of a document header stating the posting date and vendor (from the accounting view, the invoicing party and creditor), and at least one invoice document item. This item indicates which amount was charged for what quantity of a material externally procured from the vendor. Processing of the invoice receipt and the invoice verification can be done in a variety of ways:

- **Invoice verification in dialog mode**
  Invoice verification in dialog mode corresponds to the traditional procedure of invoice verification: The company receives an invoice, enters its data into the system with reference to a purchasing document, compares the recommended data and, if necessary, makes corrections. Referenced documents are generally purchase orders or individual goods receipts that are to be invoiced separately. Invoices without a purchase order reference can be directly posted to a G/L or material account. When an invoice is posted, the purchase order history is updated and information regarding payment is sent to Financial Accounting.
- **Preliminary posting of invoices**
  Invoices can be preliminarily posted by an employee. When this is done, the invoice document is entered in the system and saved without posting. Preliminarily posted documents can be edited until the actual posting is done. Only after posting can account movements and updates be performed.
- **Invoice verification in the background**
  Invoice verification in the background represents a rational processing of mass data that only needs to be manually adjusted in exceptional cases. For vendor invoices, only the allocation and total sum are recorded; the system checks the invoice automatically in the background. Documents that are not yet posted can be manually edited by an employee.

- **Electronic invoice receipt**

  For faster data transfer and to prevent typing errors during manual entry of an invoice, invoice information can be electronically transferred from the vendor to the system. Similar to invoice verification in the background, the system tries to independently post the invoices received via EDI (Electronic Data Interchange). If an invoice cannot be booked, an employee can revise it manually.

An exception to the invoice verification procedures explained above is *automatic settlements*. They are used to settle goods receipts. This process, known as ERS (*Evaluated Receipt Settlement*), is based on an agreement with the vendor that no invoice is to be generated or expected for a particular purchase order. Instead, the invoice document is automatically generated from the purchasing document data of the goods receipt. This eliminates any invoice deviations.

The basis for generation of an automatic settlement are the conditions of the purchase order and the delivery quantities actually posted. The system uses the purchase order conditions, payment conditions and tax information to determine the amount to be paid for this purchase order procedure to the external supplier (see Fig. 4.38).

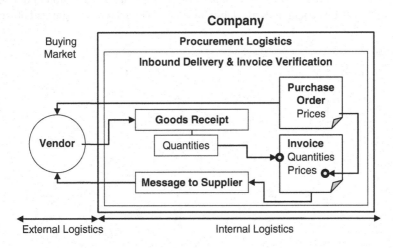

Fig. 4.38 Automatic goods receipt settlement

Another example of automatic settlement is consignment and pipeline settlement. For these types of settlements, the posted withdrawals form the basis of an internal settlement. Consignment processing is explained in Volume II, Chapter 3, "Warehouse Logistics and Inventory Management".

In addition to the net order price, the vendor invoice can also contain delivery costs. These include costs for a delivery that are charged in addition to the actual value of the delivery. We can differentiate between planned and unplanned delivery costs:

- **Planned delivery costs**
  Planned delivery costs are stipulated in advance with the vendor, a carrier or customs, and are included in the purchase order on the item level. Examples of planned delivery costs include freight or customs charges that are either indicated as a fixed amount, independent of the scope of supply, related to quantity or as a percentage of the value of delivered goods. Reserves are automatically posted for planned delivery costs at the time of goods receipt. When the invoice is received, reference is made to the delivery costs planned in the purchase order, and the reserves are compensated. The planned delivery costs are assumed in the evaluation of the stock materials during goods receipt. On the other hand, for a consumable materials purchase order assigned to an account, the account assignment object is charged.
- **Unplanned delivery costs**
  Unplanned delivery costs are unknown at the time that the order is placed and are thus only recorded in the system upon invoice receipt. Unlike planned delivery costs, there is no accumulation of reserves; instead, the material valuation made at the time of goods receipt is corrected. Such unplanned delivery costs in the invoice document can either be proportionately distributed in relation to the calculated invoice items or go to separate G/L accounts. Distribution of delivery costs to a G/L account does not debit the balances or account assignment objects.

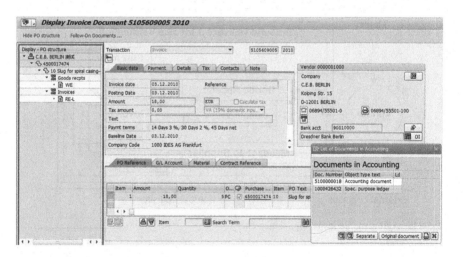

Fig. 4.39 Incoming vendor invoice

Incoming Vendor Invoice

Figure 4.39 shows the incoming vendor invoice 5105609005 for Purchase Order 4500017474. To record the vendor payment request, Accounting Document 5100000018 was generated. Figure 4.40 shows the accounting document with the posted invoice amount of Creditor 1000, the firm C.E.B. in Berlin.

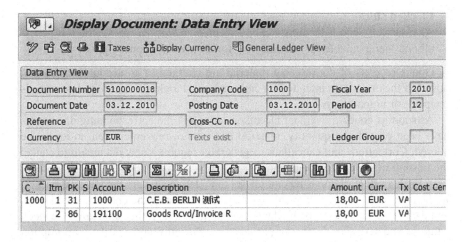

Fig. 4.40 Accounting document for incoming vendor invoice

4.5.4   Integration into Inventory Management

In the external procurement of stock material, goods receipt usually takes place into stock, for which the system increases the inventory. The procurement of consumables materials leads to goods receipt in Consumption, for which only a consumption statistic is updated in the material master.

In the following, we assume external procurement of valuated stock material. You can find a more precise description of the differences in the valuation of stock and consumable materials, as well as an explanation of the *Quality Inspection Engine* as part of SAP Extended Warehouse Management, in Volume II, Chapter 3, "Warehouse Logistics and Inventory Management".

The material document is proof that an action has caused a change in inventory. If material movement is relevant to valuation, at least one accounting document is generated in addition to the material document.

The type of inventory to be updated, the so-called *stock type*, is relevant to the determination of available stock in MRP, as well as for removals in inventory management. Goods receipt for stock materials can be posted as one of three types of stock:

- **Unrestricted-use stock**
  This type of stock makes sense if no quality inspection is to take place and there are no restrictions in use.
- **Inspection stock**
  If a quality inspection is to take place, the stock is posted to *inspection stock*. This type of stock is available for planning, but removals for consumption are not possible.

- **Blocked stock**
  Alternatively, goods receipts can be executed as *blocked stock*, which is generally neither available for planning nor for removal for consumption.

Goods receipts of consumable material are charged to an account assignment object. If the material is intended for consumption, Purchasing can stipulate a goods recipient or unloading point.

As soon as goods movement has taken place and the corresponding documents have been posted, the quantity, material and type of movement can no longer be altered. Corrections and cancellations generally require a new document to cancel the posting of the faulty document.

A further effect of goods receipt relates to updating stock in the material master: In the case of *goods receipt to stock*, the system increases the entire valuated stock and the respective stock type by the delivered quantity. At the same time, the stock value is updated. *Goods receipt for consumption* only updates the consumption statistics in the material master.

4.6   Optimizing Purchasing

It is the task of Purchasing to cover the requirements occurring in a company. In this chapter, we have already seen how these requirements are determined and in what form they are transferred to the respective department for external procurement. The reported requirement reaches Purchasing either in the form of a purchase requisition, purchase order or as a release against an existing outline agreement.

In order to satisfy the requirements determined by MRP or directly reported by the respective departments as quickly and easily as possible, purchasing procedures must be organized in an optimal manner. Automatic determination of supply sources and the management and maintenance of outline agreements contribute considerably to the rationalization potential and thus the optimization of the purchasing process.

Depending on the system employed, supply source determination can either be done in SAP ERP or SAP APO. The required master data and objects, such as purchasing info records and contracts, are transferred via the Core Interface Function (CIF) to the APO system, where they generate external procurement relationships. The process for externally procured products is done in the same way as is depicted in Fig. 4.2.

4.6.1   Supply Source Determination

In order for a purchase requisition item to be automatically converted into a purchase order, the system must know from which source and under what conditions the material is to be procured. A supply source can be an external supplier or one of the company's own plants.

Source determination in purchase requisitions and purchase orders can be based on the following objects, which will subsequently be explained in more detail:

- **Purchasing info records**
  A purchasing info record represents a relationship between the vendor and material to be procured. It contains data on a certain material and the supplier of that material, such as the current vendor price, planned delivery time and the vendor's name for the material. If a purchase order is generated with reference to an info record, the system automatically assumes its purchasing conditions in the subsequent order.

- **Outline agreements**
  If a purchase requisition has been allocated to an outline agreement, the system can generate release orders or delivery schedules.

- **Source list**
  Entries in a source list determine for a specified period of time which sources should be preferentially used.

- **Quota arrangement**
  Quota arrangements influence the allocation of possible supply sources to a purchase requisition by determining what portion of the total demand of a material may be procured from which source.

- **Plants**
  Internal procurement between locations of the same company can be performed with stock transfer orders using an internal procurement procedure. The plant of procurement represents the internal supply source.

Actual supply source determination—determining from which vendor the material to be procured is to be ordered—can either be done directly and manually when the purchase requisition is created, or automatically in the planning run.

In the case of *manual supply source determination*, the requesting party or Purchasing employee determines from which vendor a certain required item is to be procured when the purchase requisition is created.

Generally, however, Purchasing is responsible for supply source management and the allocation of sources, and is supported by the ERP system, which can recommend a list of existing supply sources. If the purchase requisition is identified as relevant to source determination (see Fig. 4.41), the system can allocate a supply source. If the system finds more than one valid supply source for an item, the decision process can be supported by a price simulation or data from the vendor evaluation.

Supply source determination can also be *automated*. Especially during requirements planning, the purchase requisition can contain not only material requirements but also the supply source that has been automatically allocated to the respective requirement item. The automatic and specific determination is supported by the *source list*.

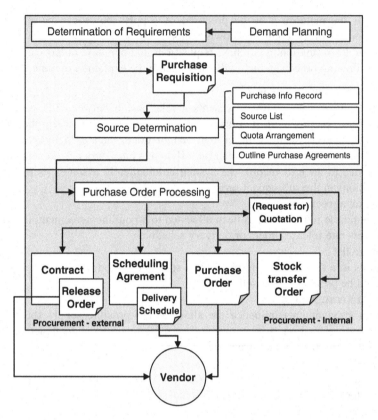

Fig. 4.41 Overview of supply source determination

Actual procurement can be conducted externally or internally. Vendors are among the external supply sources for external procurement. Other plants belonging to the same company are among the internal supply sources of internal procurement, which is done via stock transfer orders. Stock transfer orders are explained in Volume II, Chapter 3, "Warehouse Logistics and Inventory Management".

4.6.1.1   Source List

Using source lists, the possible supply sources of a material for a specific plant can be managed (see Fig. 4.42). The source list, which is taken into consideration for automatic supply source determination in Purchasing and during requirements planning, contains the supply sources of a material that are permitted and not permitted for a plant. In this regard, you can decide whether a supply source should have preference for a specified period. If nothing is allowed to be procured from a certain supplier during a specified time, this source and its source list are blocked.

We have already mentioned that the result of automatic material requirements planning can be a purchase requisition. In such a case, the supply source can be automatically determined in a purchase requisition. Automatic determination of a supply source using source list entries requires that a valid source list entry exists for the corresponding material and that it has been identified as relevant to MRP.

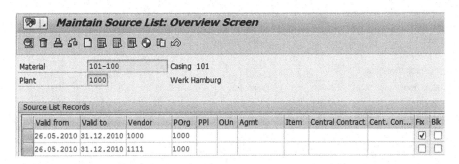

Fig. 4.42 Source list overview

Source List
Figure 4.42 shows a source list for Material 101–100 in Plant 1000. The material can be procured from May 26, 2010, to December 31, 2010, from Vendor 1000 as well as Vendor 1111. Figure 4.43 shows the detail of the source list item for Material 101–100 for Vendor 1000. Because this source list record is not relevant to MRP, it is not included in the purchase requisition for automatic supply source determination.

A *source list requirement* can be set for a material. This setting results in that material only being permitted to be procured from the supply sources cited in the source list as valid. In such a case, the existence of a source list would be compulsory. Identification of the material is done in the Purchasing view of the material master record (see also Fig. 4.11). This source list requirement can also be defined for an entire plant. For this, however, there must be source list entries with valid supply sources for all materials of a plant to be procured externally.

The source list is generally maintained per material and plant (see Fig. 4.42). The source list entries can either be made manually or automatically by the system.

- *Manual maintenance* is either performed directly or based on an existing outline agreement or purchasing info record. If an outline agreement or info record is created or edited, the corresponding contract item or purchasing information can be transferred to the source list of the material.
- For *automatic generation* of the source lists, the system offers the option of rapidly collecting all supply sources of a material in a source list and generating or updating the corresponding source list items. Automatic generation of source

lists offers a preview function with which you can simulate the effects of an automatic generation run and assess its possible consequences.

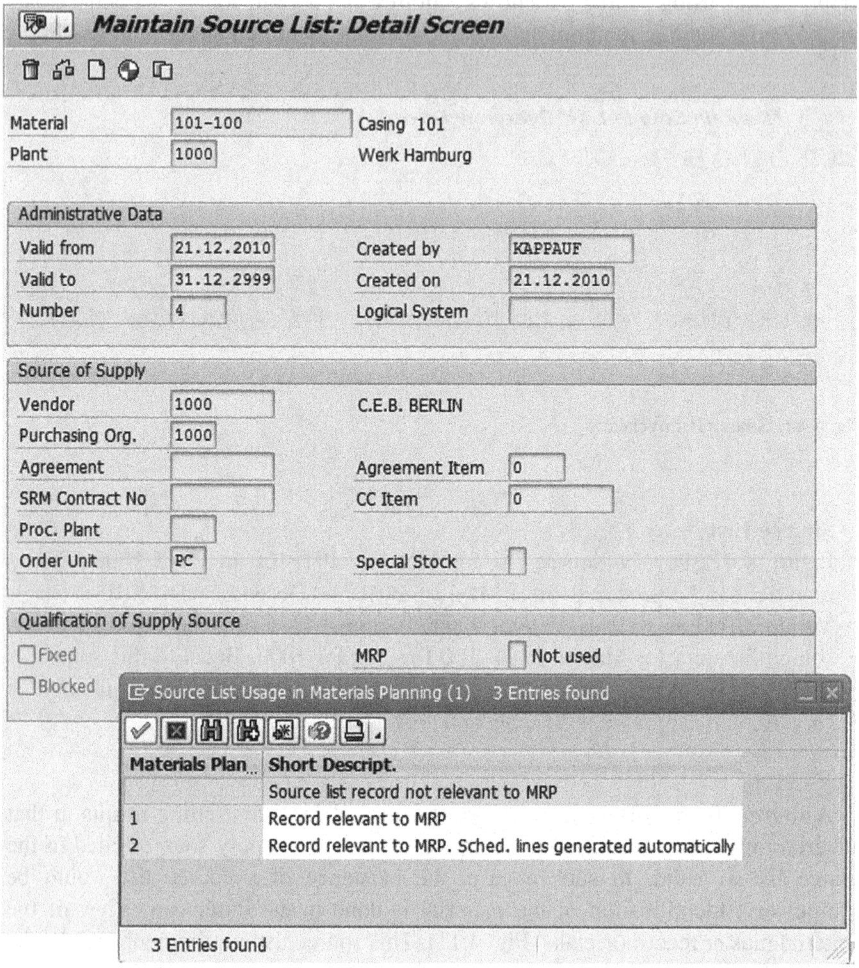

Fig. 4.43 Detail of a source list item

4.6.1.2   Quota Arrangement

Like the source list, the quota arrangement facilitates the maintenance of supply sources on the plant level. The main difference is that the individual supply sources are assigned a quota. If a certain material can be procured from various supply sources, this quota indicates which portion of the requirements can be procured within a specified period from which supply source.

As in the case of the source list, quota arrangement is used in supply source determination when a quota arrangement item exists for a particular material for the respective period of time. The allocation of a supply source using quota arrangement and the respective division of requirements among various suppliers based on a set quota is conducted automatically in the planning run. The system calculates the division percentage of requirements using the set quota and updates the quota-allocated quantity for every requirement allocation that is based on the quota arrangement. In this case, the actual requirement quantity is not divided, but rather, the entire required quantity of a purchase requisition is allocated to a supply source based on the quota arrangement.

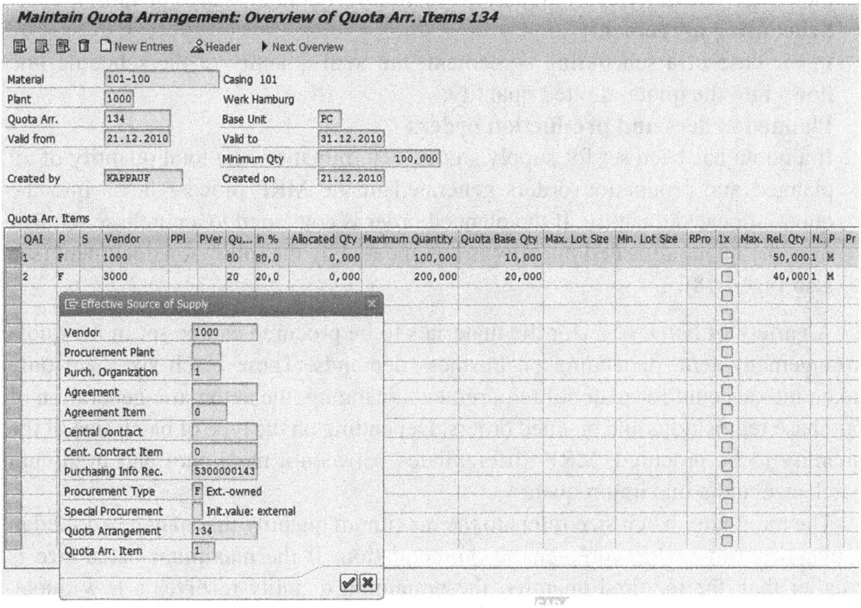

Fig. 4.44   Maintenance of the quota arrangement

Quota Arrangement
Figure 4.44 shows a quota arrangement for Material 101–100 and Plant 1000. In the period from December 21 to December 31, 2010, the material can be obtained from two vendors: 80% from Vendor 1000 and 20% from Vendor 3000. External procurement from Vendor 1000 is done based on the agreements in the existing purchasing info record 5300000143.

Supply source determination with the aid of a quota arrangement and thus the updating of the quota-allocated quantity can be done in various areas of procurement, at various times and based on a variety of purchasing documents.

When you determine a quota for supply sources in a validity period, the quota arrangement, and thus the determination of the next supply source for a requirement, can be taken into consideration in the following areas:

- **Purchase requisitions**
  Source determination in the purchase requisition process can be controlled via a quota arrangement, whereby the required quantity of a material flows directly into the quota-allocated quantity of the quota arrangement item.
- **Purchase orders**
  Orders created on the basis of a purchase requisition for which a quota has already been taken into account do not influence the quota-allocated quantity. Otherwise, the ordered quantity of a material flows into the quota-allocated quantity.
- **Scheduling agreement**
  In the case of a scheduling agreement, the total quantity of the schedule line flows into the quota-allotted quantity.
- **Planned orders and production orders**
  If a quota has been set for supply source determination, the total quantity of all planned and production orders generated in the MRP process flows into the quota-allocated quantity. If the planned order is converted to a purchase requisition, the quota-allocated quantity is not updated by the follow-on document (see also Fig. 4.18).

A variety of *batch sizes* for the materials to be procured can be set in the quota arrangement item, depending on business demands. These batch sizes are only taken into account for material requirements planning, the automatic generation of purchase requisitions and planned orders. Depending on the type of batch size of the quantity to be procured, MRP differentiates between a maximum and minimum batch size and a maximum quantity.

The maximum batch size refers to the maximum quantity that may be allotted in a planning run per procurement recommendation. If the maximum batch size is smaller than the required quantity, the remaining quantity receives a new quota. The minimum batch size, on the other hand, defines the minimum amount that must be allocated to a supply source per procurement run.

A maximum quantity can be maintained per quota allocation line. The maximum quantity is taken into consideration in the manual and automatic generation of purchase requisitions, and serves as an upper limit for the quota-allocated quantity of a supply source. The system checks whether the quota-allocated quantity is or would become larger or the same as the maximum quantity. If this is the case, that particular supply source is no longer recommended.

4.6.2 Outline Agreements

An *outline agreement* is a long-term agreement with an external supplier regarding the supply of materials or the performance of services at predetermined conditions.

The outline agreement and its stipulations apply for a defined period of time and either for a predefined total purchase quantity or a particular total purchase value.

Outline agreements generally do not contain information on delivery dates or delivery quantities. Depending on the type of outline agreement involved, the delivery date and the quantity to be delivered are either indicated in a release order or schedule line.

An outline agreement can either be a contract or a scheduling agreement. The major differences between the two types of contracts relate to the document volume and their use in automatic materials requirement planning.

Differences Between Contract and Scheduling Agreements

Contracts and scheduling agreements are outline agreements andthus long-term agreements with external suppliers. The two major differences between them consist of the following points:

- **Document volume**
  Contracts generally have a considerably higher volume of documents, since a new purchase order is generated in the system for every release order. In contrast, the scheduling agreement is only augmented by one other document, the *schedule line*, which is always supplemented with the new requirement quantities and dates.
- **Automatic planning**
  For supply source determination in material requirements planning, a contract item can be automatically allocated to a purchase requisition item as a supply source. This purchase requisition must then be converted to a purchase order, which causes the outline agreement document to be generated. The generated purchase order is called a release order. On the other hand, the scheduling agreement offers the opportunity of having a schedule line directly and automatically generated from the planning run without the creation of additional purchasing documents. (see also Fig. 4.50).

4.6.2.1 Contracts

Contracts are created manually either with or without reference to an existing purchasing document. You can thus not only create a contract without a reference or use an existing contract as a model, but you can also create contract items with regard to an existing quotation or purchase requisition.

A contract is a purchasing document whose structure and separation into a document header and document item do not differ greatly from other purchasing documents. Like other documents, the header data of a contract contains information that applies to the entire document. Such information includes vendor data, the contract period and header conditions, such as delivery costs that apply for all contract items.

The item data contains the material or service to be procured externally, prices and texts, but no exact delivery quantities or dates. The individual contract items can either refer to a single plant or all plants within a purchasing organization (see also Chap. 3, "Organizational Structures and Master Data").

For plant-independent items, we refer to what is known as a *centrally agreed contract*, whose primary advantage is that better conditions can often be negotiated for a central purchasing organization than for each individual plant. For a centrally agreed contract, the respective plant is not specified until the actual purchase order—the release order—is created. Only those plants assigned to the respective purchasing organization can release against a centrally agreed contract.

Controlling Release Orders with the Source List
Depending on business requirements, it may be necessary to block certain plants belonging to a particular purchasing organization from releasing against a centrally agreed contract. In order to prevent the centrally agreed contract from being determined as a source, it can be blocked with a source list entry for the respective plant.

Depending on whether the total quantity to be ordered within the contract period has already been reached or the total value of all possible release orders has not exceeded a specified amount, you can choose between two types of contracts: quantity contracts and value contracts.

For a *quantity contract*, the total quantity that is to be ordered within the stipulated contract period is already known, based on an agreement with the external vendor. The target quantity of material is maintained in the item data (see Fig. 4.45).

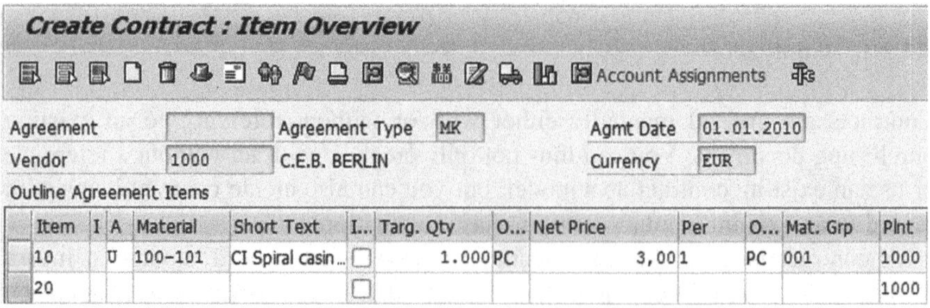

Create Contract : Item Overview

🔲 🔲 🔲 🗋 ⬜ ⬜ ⬜ ⬜ ⬜ ⬜ ⬜ ⬜ ⬜ ⬜ ⬜ ⬜ 🗏 Account Assignments ⬜

| Agreement | | Agreement Type | MK | | Agmt Date | 01.01.2010 |
| Vendor | 1000 | C.E.B. BERLIN | | | Currency | EUR |

Outline Agreement Items

Item	I	A	Material	Short Text	I...	Targ. Qty	O...	Net Price	Per	O...	Mat. Grp	Plnt
10		U	100–101	CI Spiral casin...	☐	1.000	PC	3,00	1	PC	001	1000
20					☐							1000

Fig. 4.45 Item overview of a quantity contract

> **Quantity Contract**
> Figure 4.45 shows a quantity contract stipulated with Vendor 1000. The purpose of this contract is the external procurement of 1,000 units of Material 100–101 at the conditions indicated for that item.

A quantity contract is fulfilled when the sum of all contract release orders has reached the stipulated quantity. In this regard, the so-called *release order documentation* for a contract indicates all details on order activity, the individual purchase orders with the respective quantities, as well as the cumulative total quantity and the order value. The release order documentation is automatically updated when a release order is created, and serves as the basis for contract monitoring (see Fig. 4.46).

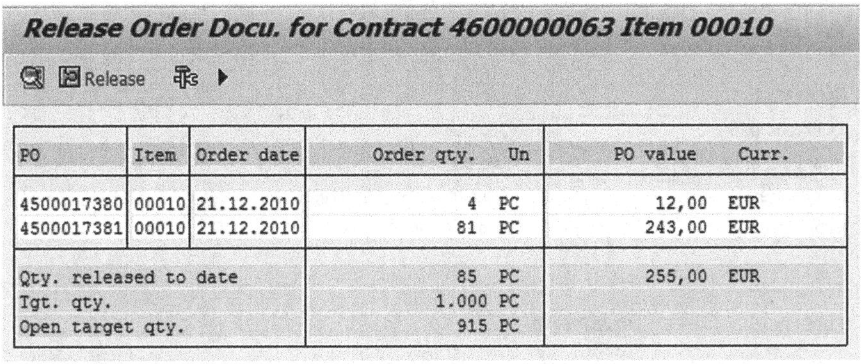

Release Order Docu. for Contract 4600000063 Item 00010

🔍 📄 Release ⟲ ▶

PO	Item	Order date	Order qty.	Un	PO value	Curr.
4500017380	00010	21.12.2010	4	PC	12,00	EUR
4500017381	00010	21.12.2010	81	PC	243,00	EUR
Qty. released to date			85	PC	255,00	EUR
Tgt. qty.			1.000	PC		
Open target qty.			915	PC		

Fig. 4.46 Release order documentation for a quantity contract

> **Release Order Documentation for a Quantity Contract**
> Figure 4.46 shows the release order documentation for Quantity Contract 4600000063 from the previous example.Of the specified target quantity, 85 units have already been released in two purchase orders. The open target quantity thus equals 915 units.

Value contracts refer to a contract type for which a total value is specified at the start that is not to be exceeded by the total release orders. A value contract is fulfilled when the sum of the release orders has reached the value specified with the vendor.

This total value, or *target value*, is maintained in the header of the value contract (see Fig. 4.47). Unlike quantity contracts, they do not necessarily have to indicate a stipulated target quantity at item level, but only the materials that are to be released against the value contract.

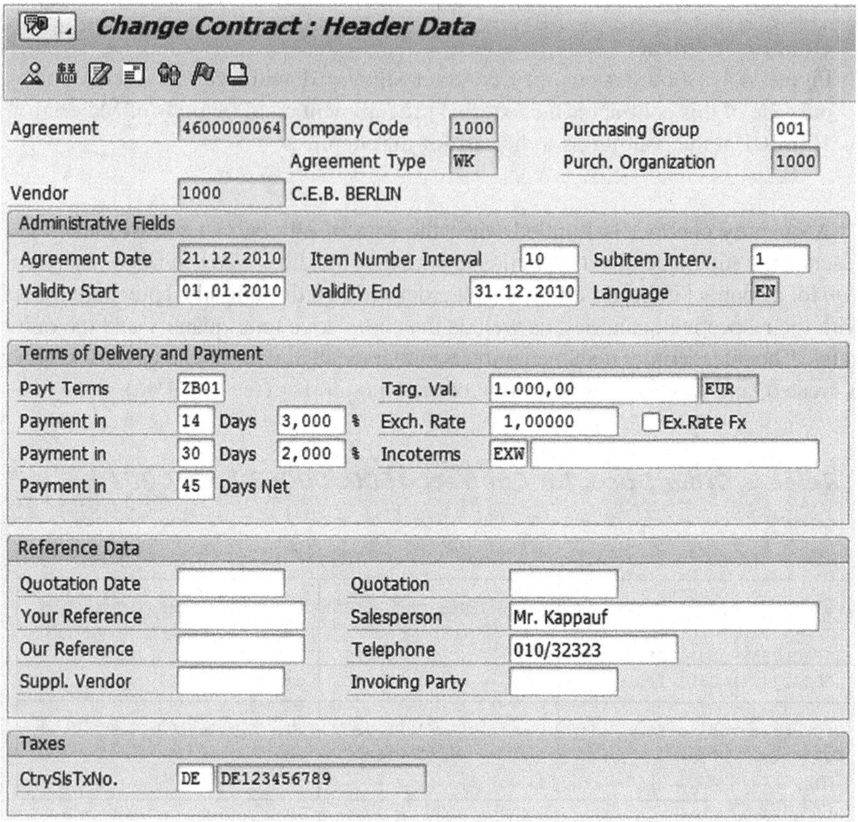

Fig. 4.47 Header data of the value contract

A special case in maintaining value contracts is the generation of contract items without a material reference. For instance, to enable procurement of office materials, creation of a contract item for that group of materials can be limited to the entry of a material group and, depending on the type of contract, the indication of a target quantity or total value.

4.6.2.2   Scheduling Agreement

Like a contract, a scheduling agreement is a longer-term agreement with a vendor, and also belongs to the group of outline agreements. The purpose of this type of agreement is the supply of materials at fixed conditions, in a specified period and at a defined total purchase quantity (see Fig. 4.48).

From a business standpoint, the scheduling agreement corresponds to a type of order having a long validity period. *Schedule lines* use these validity periods, which are separate from target quantities and conditions, to control the required partial quantities and delivery dates. These schedule lines can be created with or without reference to a purchase requisition. The external vendor is notified of the partial quantities and delivery dates upon accessing the scheduling agreement.

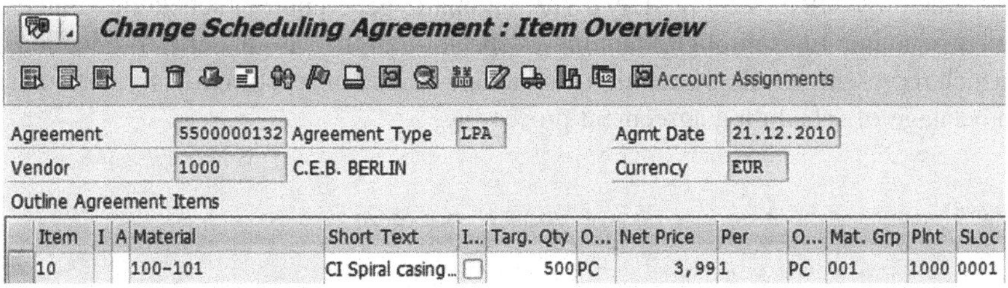

Fig. 4.48 Item overview of a scheduling agreement

Scheduling Agreement
Scheduling Agreement 5500000132 (see Fig. 4.48) is a scheduling agreement with Vendor 1000 regarding the supply of a total of 500 units of Material 100–101 at a price of €3.99 per unit. Agreement Type LPA indicates that this scheduling agreement includes release order documentation.

Scheduling agreements offer three advantages: Firstly, in contrast with purchase orders, they require less administrative effort, because they can be used in requirements planning to keep warehouse stock low using just-in-time delivery and to allow an overview of requirements over the longer term.

Secondly, scheduling agreements shorten processing times in a company, since their delivery schedule lines can replace a number of purchase orders or contract release orders. The process can be largely automated, for instance by configuring it to generate daily requirements planning and subsequent releases and perform weekly release order determination. Manual processing by the purchaser or planner would then only be necessary in exceptional cases, such as for changes in requirements within the scheduling agreement period.

Thirdly, stipulating a scheduling agreement and selecting reliable vendors shortens the procurement cycle and reduces stock, since stockkeeping is predominantly done by the vendor. In return, because of the long-term service commitment, vendors can negotiate more favorable conditions with their suppliers. These advantages can be passed on to your company in the form of more favorable procurement conditions.

Scheduling agreements are created manually. This can be done with or without reference to a purchase requisition, outline agreement request, requisition or another scheduling agreement. Scheduling agreements are always plant-specific, and can be generated with reference to a centrally agreed contract in order to take advantage of centrally agreed conditions.

Due to the close relationship between a scheduling agreement and a purchase order, the same item types are available when a scheduling agreement is created.

Scheduling agreement schedule lines contain information on the required partial quantities and delivery dates. They can either be manually or automatically generated by MRP (see Fig. 4.49). The automatic generation of scheduling agreement schedule lines through Planning—especially in the case of series production, which represents a bulk transaction with a high degree of repetition—is a particular advantage of scheduling agreement processing.

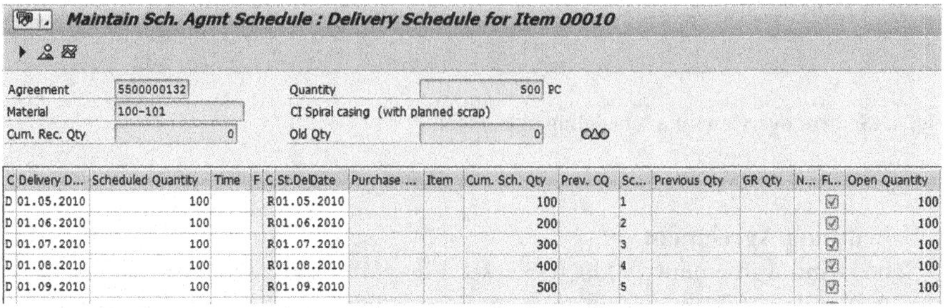

Fig. 4.49 Schedule lines of a scheduling agreement

Scheduling Agreement Schedule Lines
Figure 4.49 shows the schedule lines of a scheduling agreement. The target quantity of 500 units for Material 100–101 is divided into five quantities of 100 units each. The scheduling agreement schedule lines contain the respective schedule line quantity and the desired delivery date. The delivery date and requirement time can be entered precisely to the day, to the week or to the month.

Scheduling agreements can be used with or without release documentation. The release documentation includes data on the schedule line information transferred to the external vendor. Figure 4.50 shows the difference between processing scheduling agreements with and without release documentation.

• **Scheduling agreements without release documentation**
For scheduling agreements without release documentation, the schedule lines are transferred to the vendor in the form in which they are saved in the system. The item level of the scheduling agreement only indicates when the last release took place. Depending on system settings, the system—which does not update detailed documentation regarding the releases already communicated to the vendor—generally sends the vendor all open schedule lines for every alteration to the scheduling agreement.

- **Scheduling agreements with release documentation**
  In the case of scheduling agreements with release documentation, the schedule lines are not directly forwarded to the vendor. They serve exclusively as internal information and can only be sent to the external vendor by creating a release. The release documentation stores the transmission date, the time and the last goods receipt date, as well as the quantities and dates transmitted to the vendor, thus offering complete transparency of the release information communicated to the vendor. The release can either be a *forecast* (FRC) *delivery schedule release* or *just-in-time (JIT) release*.

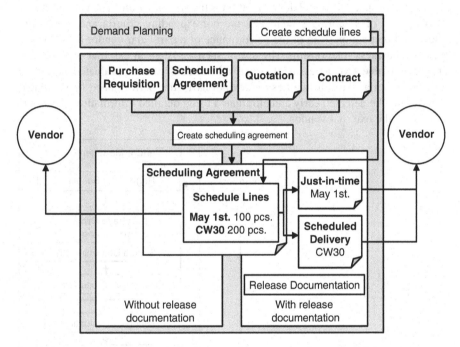

Fig. 4.50 Scheduling agreement types

The *scheduling agreement release* is only for scheduling agreements with release documentation. A release is a message to a vendor to deliver the material on the dates indicated in the schedule lines.

A scheduling agreement release can be created manually or automatically with the aid of a report, and sent as an electronic message. Automatic release generation in this case can be either performed for all selected scheduling agreement items or only for those for which schedule lines have been created or edited.

Depending on business needs, the release can be in the form of a *forecast delivery schedule release* or a *just-in-time release*. Requirement times and schedule lines in a forecast delivery schedule release are generally defined to the week or month, and provide the vendor with a medium-term overview of requirements. On the other hand, just-in-time releases show requirements to the day or even hour.

4.6.3   Vendor Evaluation

The vendor evaluation supports purchasers in selecting vendors and performing running controls of vendor relationships. As a seamlessly integrated component of Purchasing, the vendor evaluation accesses all materials and quality management data and evaluates them in accordance with your respective business needs.

The purpose of the vendor evaluation is to ensure your own competitive capability by providing Purchasing with precise information on the most advantageous prices as well as payment and delivery conditions.

A further aspect of the vendor evaluation is the *vendor rating*. The vendor rating, with the aid of a scoring system and the corresponding criteria and weighting, allows a direct comparison of the performance of a particular vendor based on the previous business relationship. The use of such a system in external procurement ensures an objective assessment, since all vendors can be evaluated based on uniform criteria. Any difficulties or deviations in the performance of stipulated services can be detected early and eliminated using detailed information in cooperation with the external vendor.

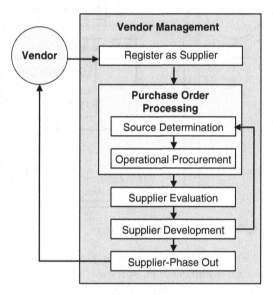

Fig. 4.51 Vendor
management

The vendor rating and evaluation are thus a central part of supply source management (see Fig. 4.51). *Vendor or supply source management* includes all business activities in the life cycle of a vendor relationship, and its goals are conducting operative procurement with the optimal supply source and further developing the contractual relationship with the respective vendor.

The evaluation of external vendors is done on the purchasing organization level, meaning every purchasing organization evaluates the vendors allocated to it. Actual evaluation is done on the basis of materials management master data, inventory management data and goods receipts with their quantities and dates, as well as based on data of the *Logistics Information System* (LIS). The LIS will be explained in Volume II, Chapter 5, "Controls and Reports".

The *main criteria* constitute the basis for vendor evaluation. Each service is rated on a scoring system from 1 to 100 points. The main criteria assessed, which ultimately determine the overall score of an external vendor and which can be defined independently by every purchasing organization, and the point scale can be altered individually (see Fig. 4.52).

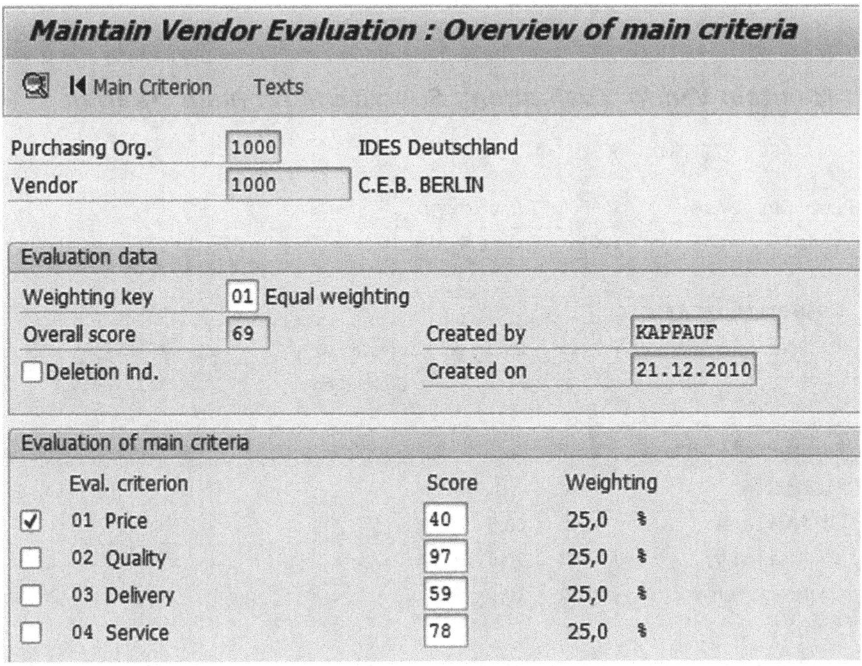

Maintain Vendor Evaluation : Overview of main criteria

🗋 I◀ Main Criterion   Texts

Purchasing Org.	1000	IDES Deutschland	
Vendor	1000	C.E.B. BERLIN	

Evaluation data

Weighting key	01 Equal weighting		
Overall score	69	Created by	KAPPAUF
☐ Deletion ind.		Created on	21.12.2010

Evaluation of main criteria

	Eval. criterion	Score	Weighting
☑	01 Price	40	25,0 ₰
☐	02 Quality	97	25,0 ₰
☐	03 Delivery	59	25,0 ₰
☐	04 Service	78	25,0 ₰

Fig. 4.52 Main criteria for vendor evaluation

Main Criteria and Overall Score
Figure 4.52 shows the vendor rating of Vendor 1000 in Purchasing Organization 1000. The four main criteria—price, quality, delivery and service—are weighted equally in this example and are evaluated individually. The evaluation of a main criterion depends on the rating of the subcriteria allocated to that main criterion. The overall score of the vendor amounts to 69 out of 100 possible points, and results from the average of cumulative main criteria points.

Depending on the criterion, the main criteria can vary in significance. That is why Purchasing can define a weighting key to control the share a criterion has when calculating the score on the next level up.

Subcriteria represent the next level in the vendor rating. For every superordinate main criterion, the system determines its evaluation based on the corresponding subcriteria. The data required for scoring of a subcriterion can be determined manually, semiautomatically or fully automatically by the system. For manual subcriteria, the scoring is entered directly by the employee. Scores for semiautomatic subcriteria can refer to existing purchasing info records. For automatically determined subcriteria, the system determines the score based on actual goods receipts and/or quality messages (see Fig. 4.53).

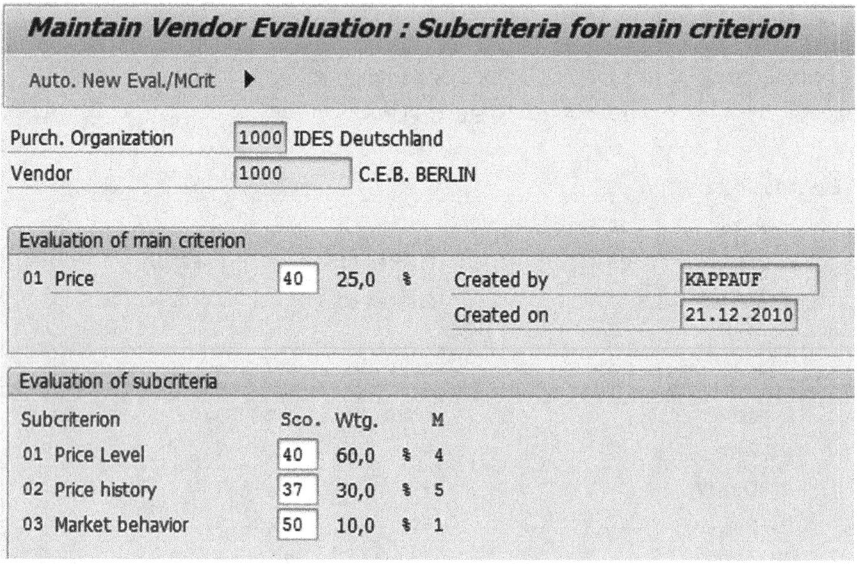

Fig. 4.53 Subcriteria of the vendor evaluation

Subcriteria and Scoring of a Main Criterion
Scoring of the subcriteria in Fig. 4.53 flows into the main criterion "Price". The main criteria score then corresponds to the average score of its subcriteria—price level, price history and market behavior. The subcriteria can generally be freely defined. However, the SAP standard system supplies the necessary subcriteria with which an external vendor can be evaluated.

The result of the vendor rating can either be displayed and evaluated in SAP ERP or within the context of a strategic supplier evaluation in SAP SRM (see Fig. 4.54). SAP ERP offers a variety of list displays and standard analyses to assess and visualize the vendor evaluation.

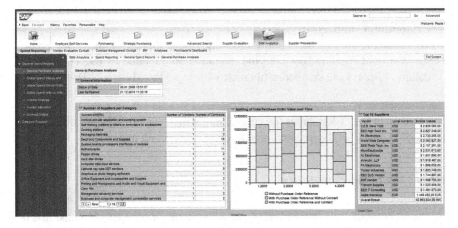

Fig. 4.54 List of the top 15 suppliers (SAP SRM)

In addition to displaying all vendors for whom evaluation has not yet been performed, you can also compare individual vendors with the average rating of all vendors from which a specific material is externally procured. The evaluation

Evaluation Comparison

⚙ Evaluation   Individual log   All logs

```
Purch. org. : 1000 IDES Deutschland
Vendor......: 1000      C.E.B. BERLIN
Eval. by      KAPPAUF      On : 21.12.2010
```

	Gen. evaluation	Material 101-100
Overall evaluations:	69	40
01 Price	40	40
01 Price Level	40	40
02 Price history	37	40
03 Market behavior	50	0
02 Quality	97	0
01 GR Lots	100	0
02 Shopfloor Complaint	100	0
03 Audit	85	0
03 Delivery	59	0
01 On-time delivery	38	0
02 Quantity reliability	84	0
03 Shipping instructs.	40	0
04 Notification reliab.	50	0
04 Service	78	0
01 Reliability	78	0

Fig. 4.55 Evaluation comparison in a vendor evaluation

Evaluation Comparison

The example in Fig. 4.55 shows the individual ratings of the main and subcriteria of Vendor 1000 and the overall score, a value of 69. The right column displays the detailed evaluation with regard to Material 101–100.

comparison in SAP ERP shows the general rating of a vendor as well as the average rating of vendors for the external procurement of a certain material.

4.6.4 Release Procedure

In several enterprises, a purchase requisition exceeding a certain value must first be approved by various departments before the actual procurement can take place. The same is true for purchase orders that fulfill specific characteristics or value limits. Approval not only ensures the control of authorizations in accordance with business requirements, but also the verification of factual accuracy, such as the indicated account assignments or supply sources.

The release procedure, as well as the purchase requisition, is supported by the following purchasing documents:

- Purchase orders
- Contracts
- Scheduling agreements
- Requests for quotation

All documents of external procurement can thus be subject to the release procedure, in which a specific document is blocked for procurement until the respective verification and electronic release have taken place (see Fig. 4.56).

The release procedure can be defined according to flexible factors. *Release conditions* determine with which release strategy a purchase requisition or purchasing document is to be released and what criteria contribute to a document being automatically released for further processing. In this context, we differentiate between release procedures with and without classification.

- **Release procedure without classification**
  The *release procedure without classification* is exclusively used for purchase requisitions whose release conditions depend on certain characteristics of the requirement item. One example would be the overall value of an item or material group to which the material to be procured belongs. The release of the purchase requisition for this type of procedure is done at item level—each item must be individually released.
- **Release procedure with classification**
  The *release procedure with classification* can be employed for purchase requisitions as well as purchasing documents. The respective conditions are defined via characteristic values and put into the system as the so-called *release strategy*. As a rule, each document field can serve as one criterion for release.

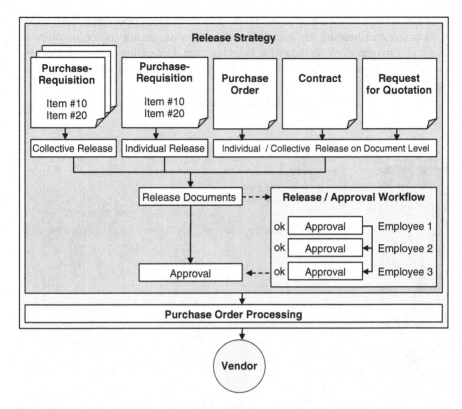

Fig. 4.56 Overview of the document release procedure

The release strategy stipulates the order of the approvals to be obtained and determines the *release codes* with which the approval is to be effected. The release code, a two-digit number that an employee can enter for approval, is dependent upon the authorizations of that employee.

The actual document release of purchase requisitions is done either as an individual or collective release. For an *individual release*, individual document items are released or rejected. A *collective release* enables the release of several purchase requisition items or entire documents. The purpose of the release of purchasing documents is to allow or prevent the printout, transmission or further processing of a document. In contrast to purchase requisitions, a release or rejection of purchasing documents is done at the header level. Item-wise release is not possible.

A message that an employee has to approve a certain document can be issued via an *approval workflow*. For this, the system generates a *work item*, a message to the respective employee in the sequence of the approvals to be issued. The work item can be displayed in a worklist (see Fig. 4.57) or directly sent to the employee via a mail system. The document to be released can be opened directly from the work item. As soon as the release has been issued, a subsequent work item can be sent to the next employee if the system has signaled that a further release is necessary.

A special type of document release is the release of a shopping cart in SAP SRM. A shopping cart holding the materials to be procured, such as office supplies, is created by an employee via the self-service function. Depending on system settings, a certain procurement value can be set that requires the approval of a superior (see Fig. 4.57).

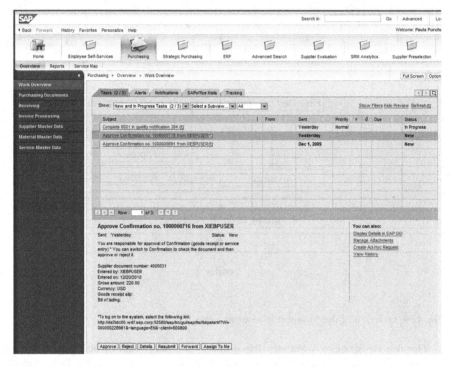

Fig. 4.57 Approval of a shopping cart in SAP SRM

Approval Workflow in SAP SRM

Figure 4.57 shows the work overview of a purchasing manager and the documents approved by him. Shopping Cart 1000000716 exceeds a set value and thus requires approval.The processing employee is informed of the shopping cart requiring approval via an approval workflow. He or she can allow the details to be displayed, approve the requirement, thus releasing it to procurement, reject it or forward it to a fellow employee.

4.7 Summary

This chapter has explained the business significance and tasks of procurement logistics and how SAP components handle them. Its objective was to present the most important systems, components and applications of SAP Business Suite and its tasks and functions with regard to external procurement of stock and consumable material, and demonstrate their integration. The following summarizes the main points:

The function of procurement is to provide an enterprise with the materials it needs for production and goods intended for sale. The materials or goods must be procured in the required quantity, type and quality at the correct point in time. The principles of cost-effectiveness must also flow into this process. Purchasing is integrated in the ERP system as part of the Materials Management (MM) component. This component supports Purchasing employees by automating several processing tasks. All documents necessary for the procurement process can be created and edited with the system. Evaluations can be generated for all purchasing-relevant activities.

The procurement cycle is typically divided into the following phases:

- **Determination of material requirements**
  The trigger for the procurement cycle is the determination of the material requirements. This can be done automatically in the SAP system or by individual departments. Requirements can either be reported directly by a department or determined by the system in the course of consumption-based planning. Section 4.3 illustrated the integration of participating systems in material requirements determination, as well as the various possibilities for the generation and source of purchase requisitions. The following chapter on production logistics explains the basics of material requirements planning with SAP ERP and SAP APO, in addition to the business significance and aspects of integration with procurement and distribution logistics.

- **External procurement**
  The basis of a procurement procedure, in addition to the required master data, includes the determination of requirements in the ERP or SCM system. The trigger for an external procurement procedure is the purchase requisition. This information, regarding what material is to be procured in what quantity, is the foundation of order processing. Based on a request for quotation issued to a vendor, this chapter introduced the purchase order process in SAP ERP: from the conversion of the purchase requisition into the actual purchase order, to supply source determination and the transmission of the purchase order to an external vendor, to the receipt of confirmations from the vendor. This chapter has addressed the differences in procuring stock and consumable materials and illustrated the significant characteristics of evaluation and account assignment. Supply source determination to cover requirements is facilitated by the evaluation of past data and existing outline agreements.

- **Purchasing**
  The processing of purchase orders, the actual purchasing procedure, can be performed on the basis of purchase requisitions, quotations and outline agreements. All documents required for procurement are generated in the system and can be transmitted electronically to the vendor. All deadlines, such as quotation or delivery deadlines, and their observance can be monitored by the system.

- **Goods receipt**
  From a materials management standpoint, a purchase order is complete when the goods have been received, and from an accounting point of view, with the receipt of the vendor invoice. Once a goods receipt has been confirmed, Inventory Management automatically updates the goods in stock. In this chapter, you have seen the significant aspects of integration with inventory management and the processing of vendor invoices. The consequences of goods movement from the standpoint of inventory management are discussed in Volume II, Chapter 3, "Warehouse Logistics and Inventory Management".

- **Invoice verification**
  Invoice verification completes the procurement process. It accesses purchase order and goods receipt data, and indicates any discrepancies in performance (quantity and price discrepancies) of the vendor. Invoice verification also serves as the basis for payment.

- **Supply source management and purchasing optimization**
  Manual and automatic supply source determination and the various outline agreements are particularly important in the optimization of purchasing, the automatic determination of supply sources and the management of agreements with external vendors. Automatic supply source determination is influenced by the functions of the source list and quota arrangements. In the realm of outline agreements, contracts and scheduling agreements were explained and the advantages of the respective document types were discussed. The evaluation and rating of vendors is part of active supply source management. The vendor evaluation offers the opportunity of securing one's own competitive advantage using materials management and quality management data. The strategic management of source supplies and their related contracts can be done in SAP SRM.

- **Release of purchase requisitions**
  In several companies, the release of purchase requisitions or other purchasing documents for external procurement depends on individual business needs—often on the purchase value and the authorizations of a particular employee. The release procedure offers the flexibility to support decisions and achieve business processes on an individual basis with the aid of an approval workflow.

The next chapter highlights the basics of production logistics and material requirements planning using SAP ERP and SAP SCM (SAP APO).

Chapter 5
Production Logistics

Production logistics is the part of the logistics chain comprising the planning and control of internal logistics processes in the realm of production. In addition to internal production logistics procedures, it includes provision planning for the required raw materials and supplies on the procurement logistics side, and transport of the finished products for distribution logistics. Figure 5.1 provides an overview.

This chapter concentrates on the functions and processes of production logistics from the viewpoint of material requirements planning. The basics of the SAP process view will be presented, along with the functional aspects of sales and procurement planning with SAP ERP and SAP APO. Production control and capacity planning with regard to actual production activity control is not within the scope of this chapter, since it lies outside the realm of logistics.

5.1 Fundamentals and Business Significance

A central function of production logistics is to supply the quantity and type of products necessary for the production processes of a company or its affiliates. In addition, the products resulting from the production process must be forwarded or disposed of according to their intended use. A significant part of this process is the monitoring, control and planning of inventories and movements in the production sites.

Production logistics centrally controls goods movement with the goal of ensuring optimal operations and thus savings.

The most important optimization aspects of production logistics are:

- Targeting improvements in sales-order-driven production with a reduction of lead times
- Increasing the flexibility of general production with an improved overview of alternative production options
- Reducing lead times in general production through timely provision of production materials, a reduction of internal production logistics procedures (transport between production stages) and a timely removal of produced products

J. Kappauf et al., *Logistic Core Operations with SAP*,
DOI 10.1007/978-3-642-18204-4_5, © Springer-Verlag Berlin Heidelberg 2011

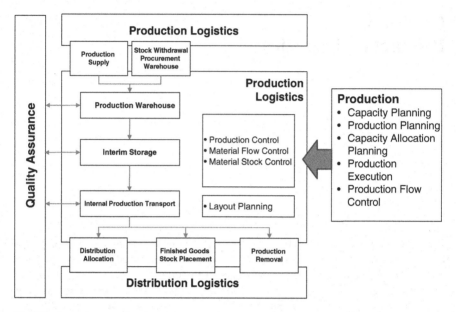

Fig. 5.1 Operations and integration of production logistics

- Reducing inventory through proper planning of the required production materials at the respective sites, driven by medium-term demand planning
- Optimizing transport routes within the production areas and between production stages
- Reducing the number of production result variants and the diversity of production materials
- Coordinating production batch sizes with internal transport and warehouse logistics
- Sensibly combining in-house production and external procurement (make or buy)

A significant goal in all of these aspects is the reduction in overall production costs. This goal can be examined from a strategic, tactical and operative perspective:

- *Strategic planning* refers to planning for the entire company over the long term. Examples of strategic production logistics tasks include the planning of production locations, suitable layouts for production plants, and make-or-buy decisions for products or components.
- *Tactical planning* optimizes processes in a localized area or in the medium term. One example of a tactical task is selecting a suitable production component supplier.
- *Operative planning* is concerned with the individual steps within production logistics that have a direct influence on production. An example of an operative task is the timely activation of transport jobs or the selection of a suitable transport means.

The following section provides an overview of the systems and components that you can use within SAP Business Suite for the realm of production logistics.

5.2   SAP Systems and Components

SAP Business Suite, with SAP ERP and SAP SCM (SAP Supply Chain Management), has several components that support the various areas of production logistics. Figure 5.2 provides an overview of them.

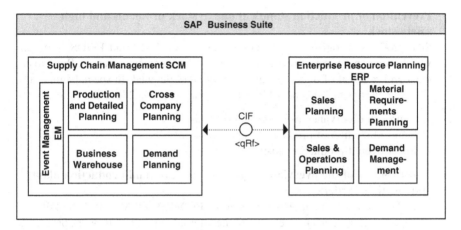

Fig. 5.2 Components of SAP Business Suite in the realm of production logistics

The use of SAP Business Suite components enables producing companies to achieve process improvements in several ways.

In the area of *demand planning*, there is often potential to improve on the following:

- **Material requirements adapted to changing customer needs, preventing quantity shortfalls or lost sales**
  The SAP solution for production logistics can contribute to the integration of supply chain functions with neighboring logistics areas through better adaption to changing markets and customer needs. A multi-level forecasting function allows you to make forecasts for a variety of periods. This will enable you to achieve the following:

  - fewer quantity shortfalls
  - shorter contract fulfillment lead times
  - increased percentage of "perfect" orders
  - fewer lost orders because of longer delivery periods or the unavailability of products

- **Improved requirements forecasting and reduction of capital tied up in unsold goods in stock**

Due to the high degree of integration of planning and execution, you can achieve greater planning efficiency and a better ability to solve conflicts. Active system forecasts support improved cooperation between planning and operative business. This will lead to:

– better forecasting accuracy for material requirement determination
– lower stock levels and thus lower stock costs

In the realm of the *production warehouse* and internal production transport itineraries, the SAP solution offers you the following improvement potential:

• **Prevention of production standstill due to stock problems and high safety stock**
SAP production logistics, with its components PP-MRP and PP/DS, supports you through an optimization of internal goods movement processes. Replenishment and removal of material can be definitively planned. In special situations, ad-hoc activities are also possible. This enables you to achieve the following:

– improved observance of the production plan
– better utilization of production capacity
– lower overall production costs

• **Improvement of the overview of goods in stock beyond production plant and location borders**
With SAP production logistics (PP-SOP, PP-MRP, PP-MP, PP/DS, DP and SNP), you will receive an overview of the internal requirements, work processes and available or procurable stock quantities. You will also obtain a clear depiction of the logistics structure and material flow within your company. This will enable you to achieve the following:

– improved opportunities for internal procurement of required production material
– improved human resources planning with lean production materials
– cost-optimized planning of goods movement within a company affiliation network

In the realm of *distribution logistics*, production logistics flows seamlessly into the areas of distribution, warehouse management and transport management. Here, too, SAP Business Suite offers successful integration of production and order-based process steps with the relevant delivery processes.

The components described below form a uniform, continuous process to treat all requirements with flexible planning and execution strategies tailored to the situation.

The following production logistics components are available in SAP ERP, which will be explained in more detail in the coming sections:

- **Sales & Operations Planning (PP-SOP)**
  Sales & Operations Planning is a forecasting and planning tool with which you can plan goals in the realms of sales and production for the medium and long term.
- **Demand Management (PP-MP)**
  The task of Demand Management is the determination of requirement quantities and delivery dates for finished product assemblies.
- **Material Requirements Planning (PP-MRP)**
  The central task of Material Requirements Planning is to secure material availability, meaning the procurement of required quantities for internal and sales purposes in a timely manner.

SAP Advanced Planner & Optimizer (SAP APO) offers a completely integrated palette of functions that you can use to plan and execute your production logistics processes:

- **Demand Planning (DP)**
  You can use Demand Planning to generate forecasts for the demand of your company's products on the market. These components enable you to consider the many factors that influence demand. The result of demand planning is the demand plan.
- **Supply Network Planning (SNP)**
  Supply Network Planning integrates procurement, production, distribution and transport, allowing the simulation and implementation of comprehensive tactical planning decisions and supply-source decisions based on a global and consistent model.
- **Production Planning and Detailed Scheduling (PP/DS)**
  Production Planning and Detailed Scheduling can be used to generate procurement recommendations for internal production requirements or external procurement, as well as to plan in detail and optimize resource allocation and order dates.

These components (see also Fig. 5.2) form the foundation of SAP production logistics processes. The *Core Interface* (CIF) guarantees seamless integration of ERP and APO. To include analytical data, such as with regard to forecasts, Business Warehouse (BW) is used, and is directly integrated with SAP APO. Another component that interacts closely with production logistics is the global availability check (*global Available-to-Promise*, gATP), which will be explained in detail in connection with distribution logistics.

5.3   The Fundamentals of Production Logistics with SAP ERP

Production logistics in SAP ERP includes procurement planning with the subareas of sales and operations planning, demand management and material requirements planning, which will be explained in more detail in subsequent sections. Figure 5.3 provides an overview of how the subareas interlock and how they are integrated with other ERP components (Production, Controlling).

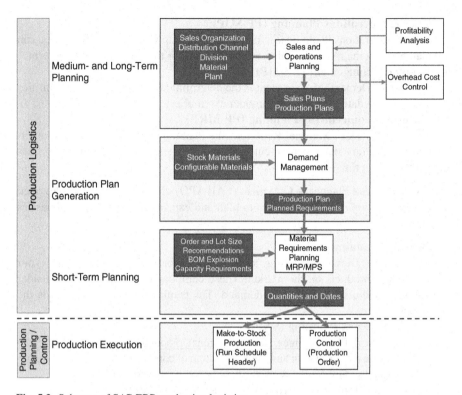

Fig. 5.3 Subareas of SAP ERP production logistics

Sales & Operations Planning (SOP) is a planning tool built on the Logistics Information System (LIS, see Volume II, Chapter 5, "Controlling and Reporting"). It transfers sales and production plans to Demand Management, which generates a demand plan and planned requirements. The demand plan and planned requirements are then forwarded to Material Requirements Planning (MRP) and Master Production Scheduling (MRP), where precise quantities and dates for production are determined.

The generation of make-to-stock and production orders for production control builds upon quantity and date targets, yet is not part of production logistics. Rather, it is part of production planning and control.

In addition to LIS data, which serves as the basis of sales and operations planning, you can use data from the profitability analysis (CO-PA) for demand planning. You can also transfer planning results to profitability analysis, cost center accounting and activity-based costing.

5.3.1 Sales & Operations Planning

Sales & Operations Planning is a logistics planning tool in SAP ERP with which you can plan target quantities for sales and production on the medium and long term. Planning is done on the basis of historical data, current key process and stock data, and standard values for the future. The historical data is taken from the Logistics Information System, where sales, production and stock data are stored for purchasing organizations, distribution channels, divisions, materials and plants (see also Volume II, Chapter 5). The generated plan not only contains data on the availability of produced goods, but also indicates the respective resources and capacities required.

Sales & Operations Planning includes two application components:

- **Standard Sales & Operations Planning (also known as Standard SOP)**
  The ERP system generates the sales and production plans according to largely fixed procedures. The results are depicted in standardized form.
- **Flexible Planning**
  With Flexible Planning, you have the option of tailoring the system to your needs using a variety of parameters. You can execute planning on every organizational level (such as the sales organization, material group, production plant, product group, material level, or from the view of the entire company) and configure the display of the results according to your specifications with regard to content and layout. This can be done via a planning table, which is similar to a spreadsheet table. In addition to accessing previous data or forecasting future market demands, you can also conduct analyses and what-if simulations.
  Due to its high level of efficiency, Flexible Planning is the more important of the two components.

Further Components for Demand Planning
Within the context of SAP Business Suite, Sales & Operations Planning is only one possibility to conduct preliminary planning for production. A further option is offered by SAP Advanced Planner & Optimizer (SAP APO) with its component Demand Planning (DP), which will be explained in Sect. 5.4.1.

The decision whether to use both components for demand planning depends on the required range of functions, and on software licensing and the business costs of running a multiple-system landscape.

As mentioned above, Sales & Operations Planning is based on *LIS structures*, which contain historical and current data on key logistics figures pertaining to organizational and master data structure. Planning data (future planned data) can also be saved in the LIS.

A *planning hierarchy* depicts the organizational levels and units of your company in a form relevant to planning. You can define a planning hierarchy from a combination of values based on the characteristics of an information structure. A planning hierarchy must exist before planning is executed if you wish to employ consistent and level-by-level planning. Figure 5.4 shows an example of a planning hierarchy.

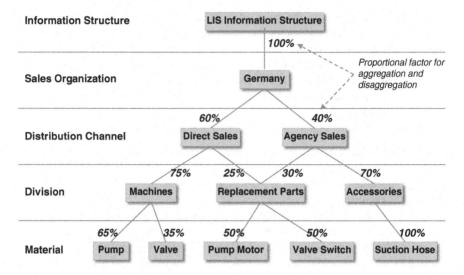

Fig. 5.4 Example of a planning hierarchy

Using aggregation and disaggregation, you have the opportunity to conduct weighted planning over multiple-level planning hierarchies.

- *Aggregation* refers to planning on an aggregate level, in which the planning data of a lower level is aggregated on the next level up. For instance, aggregation can be performed using a totaling of characteristics weighted with proportional factors. Further aggregation types include the generation of averages or the simple assumption of data.
- For *disaggregation*, the values from a higher, aggregated level are broken down on a lower, detail level using allocation algorithms and proportional factors. For this, a decision scheme is defined in the system that, depending on the aggregation type and data and/or edit status, decides how disaggregation will take place.

The *planning method* is a central element of Sales & Operations Planning that is used on the information structures of the LIS. It determines which data is to be distributed to various locations of a company. Two different planning methods are available:

- Consistent planning
- Level-by-level planning

For *consistent planning*, the planning data is stored on the planning hierarchy level with the highest degree of detail. This means that for every planning project, you always have disaggregation according to individual materials, plants or sub-organizations. Changes in plans that you perform on one level have a direct effect on the data of all other planning levels. This translation of data through aggregation and disaggregation is performed automatically. One advantage to consistent planning is its ease of use: By entering planning figures on a single level, you automatically achieve data consistency on all other levels. A further advantage is being able to create planning data under user-defined aspects.

In *level-by-level planning*, the data is stored on all planning levels. You can plan every level of the planning hierarchy individually; the various planning levels are independent of one another. This can lead to the plans no longer being consistent on various levels.

Level-by-level planning enables you to perform top-down as well as bottom-up planning. In both cases, it is possible to have the system automatically determine the proportional factors of characteristic values based on existing data (historical or planning data). This gives you the option of initially beginning a new planning project on the basis of the previous year's data, for example, and then adapting it to new or altered influences and factors. The advantage of level-by-level planning is that you can individually edit and check data before you aggregate or disaggregate it on the higher or lower levels.

If you use consistent planning, you can generate planning hierarchies using a master data generator. In the case of a standard SOP, you can use a report that can generate a planning hierarchy based on the planning data of an information structure. Of course, you can also manually create planning hierarchies. The maintenance of planning hierarchies is done from level to level, and you can determine the aggregation and proportional factors for individual characteristic values.

With the aid of SAP ERP, the Forecast function enables you to estimate the development of figures in a future-oriented time series using historical values. In Standard Sales & Operations Planning, you can forecast sales volumes for product groups or materials.

For instance, if you create a forecast for a material, its previous consumption values are analyzed, and using a suitable forecast model, a prediction can be made about future material consumption. This process includes all types of consumption data, even that which has been posted as scrap material.

In flexible planning, you can forecast any desired key figure, as long as the respective key figure is provided in the system's Customizing function. In level-by-level planning, you can also conduct a forecast for new materials with the aid of the consumption quantities of a *reference material*. This is necessary when no historical values exist for a particular material.

When analyzing key figure time series, you can determine a variety of phenomena that are defined in Sales & Operations Planning as forecast models (see Fig. 5.5):

- **Constant model**
  The time series development of a key figure statistically fluctuates by a certain average value. Constant models can, for instance, occur in a well-saturated market and an established product.
- **Trend model**
  The key figure value fluctuates statistically, yet rises or falls according to a trend over an extended period of time. Trend models can occur for a new product that gradually becomes accepted by the market and is then in demand.

- **Seasonal model**
  The key figure value fluctuates statistically as well as seasonally, so repetitions of key figure value development occur periodically. Seasonal models can be used for seasonal goods (such as snow tires).
- **Seasonal trend model**
  For seasonal trend time series development, seasonal fluctuations occur in a continuously rising average. This model can be observed in an unsaturated market for a product whose popularity is on the rise (such as a delicious new flavor of ice cream).
- **Copy actual data**
  This model only executes a transfer of actual key figure values to the forecast period (a forecast per se is not conducted).
- **Irregularity**
  If no regular pattern can be identified for a key figure series besides statistical scattering, a forecast is not possible due to the irregularity.

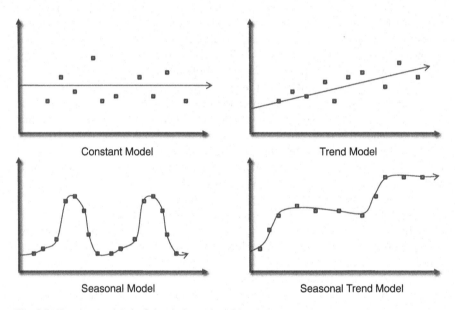

Fig. 5.5 Forecast models in Sales & Operations Planning

If planning is to be conducted solely based on historical values, the automatic forecast function offers a good basis for future key figure or consumption flows. In many cases, however, events occur that considerably influence the sales or consumption flow. Such events can, to some extent, be planned in advance and integrated into a forecast. The following are examples of such events:

- Changes in price, such as special offers that enable increased sales
- Special agreements with major customers or vendors from or to whom large quantities are bought or sold at a lower price

- Delivery problems of competitors that lead to increased customer demand
- Market information that opens up opportunities to tap new markets

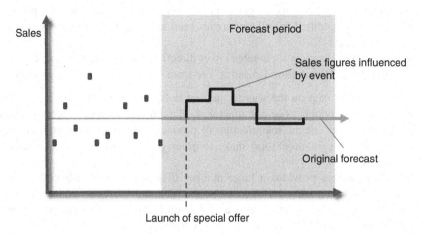

Fig. 5.6   Effects of events on a forecast with a constant model

Sales & Operations Planning allows you to integrate the effects of events into forecasts if said effects cannot be derived from historical data. In Fig. 5.6, you can see the influence of an event on a forecast generated with a constant model. Such a consumption or sales flow can be the result of a special offer leading to increased sales ("hoarding"), followed by a drop in sales due to short-term market saturation.

The planning data determined in Sales & Operations Planning can subsequently be transferred to Demand Management, which determines the requirement dates and quantities for production as well as strategies for production or external procurement of finished products.

5.3.2   Demand Management

Demand Management represents the interface between sales-oriented Sales & Operations Planning and detailed production planning, in which precise material requirements on the parts level (MRP) and the exact production steps are determined (see also Fig. 5.3). The core function of demand management is the determination of requirements quantities and delivery dates for finished products.

What Is a Requirement?
A *requirement* in SAP terminology refers to the required quantity of a material at a certain time in a specific plant.

Demand Management utilizes two types of *independent requirements* as input values for the generation of the demand program:

- Independent requirements are defined as the requirement quantities of a product planned for a specified period that are not related to a sales order (such as safety stock, stocking)
- Customer requirements are created either directly from the items of a sales order (see Chap. 6, "Distribution Logistics") or from sales scheduling agreements.

An elementary step on the way to the demand program is defining a planning strategy for the products to be manufactured. Planning strategies are economically prudent procedures for the manufacture of products. There are planning strategy variations for make-to-stock and make-to-order production, as well as external procurement.

The SAP system provides a large number of production planning strategies. Figure 5.7 shows an overview of a few of the planning strategies defined in SAP ERP Customizing. By using these strategies, you can decide how production and any necessary assembly mounting are to be conducted:

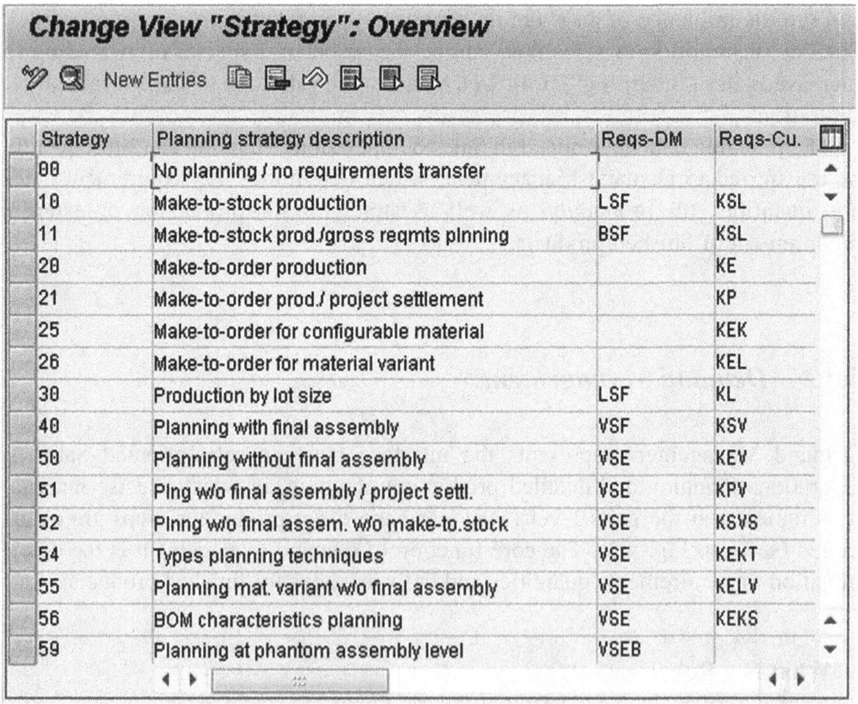

Change View "Strategy": Overview

🖉 🕄 New Entries 📭 🗟 ⬿ 🗟 🗟 🗟

Strategy	Planning strategy description	Reqs-DM	Reqs-Cu.
00	No planning / no requirements transfer		
10	Make-to-stock production	LSF	KSL
11	Make-to-stock prod./gross reqmts plnning	BSF	KSL
20	Make-to-order production		KE
21	Make-to-order prod./ project settlement		KP
25	Make-to-order for configurable material		KEK
26	Make-to-order for material variant		KEL
30	Production by lot size	LSF	KL
40	Planning with final assembly	VSF	KSV
50	Planning without final assembly	VSE	KEV
51	Plng w/o final assembly / project settl.	VSE	KPV
52	Plnng w/o final assem. w/o make-to.stock	VSE	KSVS
54	Types planning techniques	VSE	KEKT
55	Planning mat. variant w/o final assembly	VSE	KELV
56	BOM characteristics planning	VSE	KEKS
59	Planning at phantom assembly level	VSEB	

Fig. 5.7 Definition of planning strategies in demand management

- **Production triggered by sales orders (make-to-order production)**

  - make-to-order production: Requirements can be directly converted to production orders.
  - preliminary planning without final assembly: customer-specific components are preliminarily produced, and assembly is performed when the sales order has been received.

- **Make-to-stock production without concrete sales orders**

  - anonymous make-to-stock production
  - batch production
  - preliminary planning with final assembly
  - preliminary planning on the assembly level

- **External processing (service orders)**

  - Production is performed externally by another company.

Make-to-stock production strategies use data from Sales & Operations Planning to carry out the production planning. Requirements stemming from sales orders are fulfilled from stock in this case. Make-to-stock strategies are especially useful where sales levels fluctuate but production facilities need to be used to capacity for reasons of efficiency and cost.

The strategy you select will define the character of the individual phases of the demand program and the adjustment of independent and customer requirements, such as:

- The generation of the demand plan using sales orders and/or sales forecast values
- Shifting of the stocking level down to the assembly level, such that final assembly is triggered by the receipt of the sales order
- Execution of the demand plan especially for the assembly

In the case of hierarchical products, planning strategies can be combined. This enables you to select the planning strategy *Planning with Final Assembly* for a finished product, and for an important assembly in the bill of material (BOM) of that finished product, to conduct *Planning at Assembly Level*.

Determination of the planning strategy for a material to be produced can be done via an allocated strategy group. In addition, various requirement types are defined for every strategy that especially control the process of requirements allocation on the independent and customer requirements side.

The demand plan is carried out by determining planned requirements, or more precisely, planned independent requirements and customer requirements. A planned independent requirement is determined either from a planned quantity and date, or from several planned independent requirement schedule lines if the total planned quantity is divided among several production dates. Customer requirements represent the production-specific view of the requirements of a sales order. (For details on sales orders, see Chap. 6, "Distribution Logistics".) Planned requirements and the production plan are determined in the system from demand plans and sales orders using defined product requirements. The system uses the planning strategies allocated to the respective materials for this purpose. The generated production plans are subsequently made available for material requirements planning.

5.3.3   Material Requirements Planning

The main purpose of *Material Requirements Planning* (MRP) is to secure the availability of materials for production and assembly in quantities that allow all planned production procedures as well as all sales processes to be executed in a timely and accurate manner. Important subfunctions include:

- Monitoring of material stock for Production and Sales and Distribution
- Generation of procurement recommendations for Purchasing and Manufacturing
- Determination of the optimal balance between the best possible service level and lowest possible capital lockup, coupled with low provision costs, in order to reduce expenses

Figure 5.8 shows the principle process of material requirements planning and its integration in production.

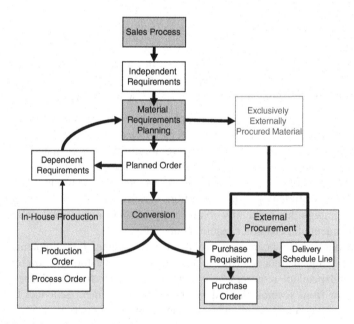

Fig. 5.8 Material requirements planning process

In the realm of material requirements planning, employees' daily tasks include the correct provision of materials according to type, quantity, location and date, and the prerequisite determination of fulfillment of demand. The following input parameters are required for this determination:

- Independent requirement figures from the sales process or demand management
- Dependent requirement figures from a so-called bill of material (BOM) explosion of products in production
- Current stock figures for all relevant materials
- Stock reservations with scheduling for all relevant materials
- Stock quantities in the ordering process with confirmed delivery dates
- Procurement lead times for all materials that can be procured externally
- Production flow times for all materials requiring in-house production

Based on this data, the MRP system can generate a procurement recommendation for the planning employee, which either leads to a planned order or to the creation of a purchase requisition for external procurement (see also Chap. 4, "Procurement Logistics").

Requirements planning can be done in the planning run on a multi-level basis, which leads to the dependent requirements cited above. In the example shown in Fig. 5.9, you can see the material requirements planning for the material "blender", which is exploded via the BOM into the components "blender drive" and "blender housing". The blender drive, in turn, is composed of a motor and a gear box.

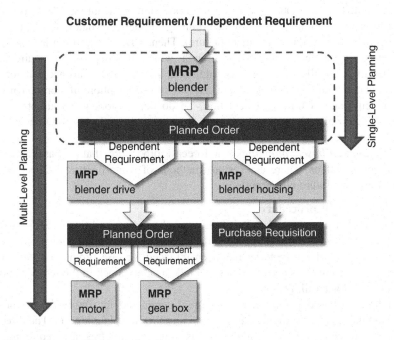

Fig. 5.9 Single- and multi-level planning

All parts except the housing are manufactured in-house (planned order), and the housing is externally procured (purchase requisition). A planning run can flow through all levels of the material hierarchy and execute a total planning process.

A number of MRP procedures are available for material requirements planning. The *MRP procedures*, which are either deterministic or consumption-based, serve to align existing stock, capacities and confirmed or expected stock additions with the required quantities. A special MRP procedure for this purpose is *Master Production Scheduling* (MPS),, which can be employed for the particularly important materials of a company. *Master schedule items* are those materials that influence the value creation of an enterprise to a high degree.

The objectives of *deterministic material requirements planning* result from the requirements of current and future sales:

- Customer orders
- Planned independent requirements
- Material reservations
- Dependent requirements obtained through BOM explosion

With a deterministic MRP, you can generally work with low safety stock, since the exact requirement quantities are known. In the first step, the MRP system conducts a net requirements calculation, in which requirements are offset by the available stock and planned stock additions. If the available stock including additions is smaller than the requirement quantity, the respective procurement recommendations are generated. Thereafter, the optimal batch size for every material is determined for the purchase order or production. Scheduling of the procurement recommendations is done by determining the delivery dates (for a purchase order) or production dates (for manufacture). Then, a BOM explosion is conducted for internally produced materials, through which the dependent requirements are identified. Finally, you can take additional requirements into account for the deterministic planning process, such as excess consumption of components in Production, using a forecast calculation. To do this, historical consumption data is consulted, which demonstrates how high excess material consumption was for certain production runs.

Consumption-based planning is a forecast procedure that is based on past consumption values. There are various methods of using historical consumption data to forecast future consumption:

- **Reorder point procedure**
  Procurement is triggered when the sum of plant stock and fixed additions falls under the *reorder point*. The reorder point must be high enough to cover the period until the material has been replaced, or else production or supply shortfalls result. It is a good idea to include a safety stock in the reorder point.
- **Forecast-based planning**
  In forecast-based planning, a forecast is conducted for future requirements. These values then form the requirement values for the planning run. The forecast values directly influence requirements planning as forecast requirements.

Forecast calculation is done at regular intervals, which enables the requirement to be adjusted to consumption behavior. Material consumption diminishes the forecast requirement, which prevents it from again being integrated into the planning.

* **Time-phased planning**
  If a material is supplied by a vendor at regular intervals (common in retail processes, such as a delivery every Wednesday), planning should be adjusted accordingly. The planning process should then take place at a time that is n days before the delivery date, whereby the variable "n" represents the delivery period. However, you can also manually bring the planning date forward. Requirements planning can be conducted at the scheduled time either in a consumption-based or deterministic manner. The requirements relevant to time-phased planning flow into the net requirements calculation in such a case.

For time-phased planning, the system additionally employs a range of coverage profile, indicating the minimum, target and maximum safety stock, in order to be able to adjust the procurement quantities. The system always tries to have stock to match the target safety stock amount at the end of the planning period.

When planning master schedule items that must be explicitly identified as such in the material master, special care is advised, since these materials are of such strategic significance to a company. Requirements planning with master production scheduling is deterministic. It is of prime importance to increase planning stability while preventing excessively high capital lockup through too much stock. Because of the importance of master schedule items, a company may tend to establish safety stock in excessive amounts, which, especially in the case of expensive materials, leads to high costs. The planning of master schedule items is always done in a separate planning run, meaning master schedule items are never planned within the deterministic or consumption-based planning process. Master production scheduling contains only master schedule items. For levels below the master schedule, dependent requirements are generated but not planned. After master production scheduling and its interactive verification, you can then start further planning runs to include dependent parts in the planning.

When a planning run is triggered, the system executes several process steps:

1. The first process step in MRP is checking the material requirements planning file, which controls and notes which materials are planned in the various types of planning runs. Generally, all materials relevant to a planning run are stored in the material requirements planning file. If the check finds an alteration to a material relevant to planning, that material is planned in accordance with the planning instructions. Any existing procurement recommendations are processed according to the instructions in the planning file.
2. A net requirements calculation is performed for every material. Available stock as well as the fixed planned additions from Purchasing or Production flow into the net requirements calculation. If the requirement cannot be satisfied with this stock, a procurement recommendation is generated, in which any defined range of coverage profiles is also taken into consideration.

3. The procurement quantity calculation is performed, taking into account set batch sizes and additional scrap quantities.
4. The start and end dates of the procurement recommendations are calculated.
5. The procurement recommendations are used to generate either planned orders, purchase requisitions or schedule lines. When the necessary information has been entered for the procurement quota arrangement, the vendor is also determined and the procurement recommendation directly allocated.
6. For every procurement recommendation of an assembly, the BOM is exploded and dependent requirements are determined (see Fig. 5.9).
7. If critical situations occur during the planning run that require manual action by a planning employee, exception messages are generated and, if necessary, a rescheduling check is performed. In addition, detailed range of coverage data is determined.

Normally, requirements planning is conducted per plant, but it can also be performed on the level of location, planning area or as supply network planning. Planning procedures include:

• Total planning
• Single-level individual planning
• Multi-level individual planning
• Interactive planning
• Multi-level individual customer planning
• Individual project planning

To check the generated procurement recommendations, a planner has a variety of evaluation tools:

• MRP list
• Current requirements/stock list
• Planning result (corresponds to the MRP list with an individual evaluation layout)
• Planning situation (corresponds to the requirements/stock list with an individual evaluation layout)
• Planning table for make-to-stock production

After the final check has been completed, the purchase requisitions can be released to Procurement Logistics and the planned orders to Production.

5.4 The Fundamentals of Procurement Planning with SAP APO

SAP Advanced Planner & Optimizer (SAP APO) offers a wide range of logistics planning tools that are supported by several optimization tools. Via the Core Interface (CIF), SAP APO is closely integrated with the procurement, sales and distribution and production processes in SAP ERP.

SAP ERP already offers a complete set of production logistics applications. So why, one might ask, would similar applications be developed in a new system, ostensibly "in competition" with the existing one?

The second half of the 1990s saw an increasing demand from major enterprises for software optimization tools. In addition, there was a tendency toward multiple-system ERP landscapes for logistics processes within corporations, based on several reasons (relating to load sharing, autonomy, decision-making authority, etc.). In other words, corporations were using one ERP system as a global platform, yet each national organization operated its own ERP system. Total consolidation was only taking place on a financial level for the consolidated balance sheet. However, because the separation of those systems proved disadvantageous for logistics processes and synergies were being lost, a central planning platform was conceived for logistics (see Fig. 5.10), from which SAP APO was developed.

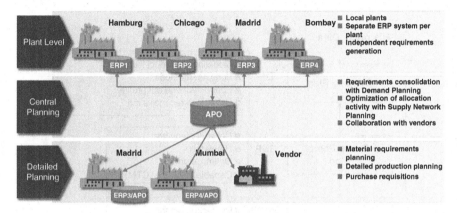

Fig. 5.10 Central planning in a network of several ESP systems

Central and Local Planning with ERP and APO
Figure 5.10 shows a typical application scenario that can no longer be managed with ERP production logistics alone.

A corporation in the chemical industry operates plants in four global locations. Each plant operates its own ERP system, in which the independent requirements for production are generated. Since intermediate products are produced in various plants and then distributed in the network, a central planning system (Demand Planning) including allocation planning (Supply Network Planning) is conducted. However, this cannot be performed in one of the ERP systems, because only the plant and material data of that particular location are known to the system. Therefore, planning must occur on a central level, which is provided by a central APO installation. Here, planning and optimizing can be done on a cross-plant basis and throughout the entire logistics network.

After this "master plan" has been generated, detailed instructions are communicated back to the plant level. In addition to the ERP system, each plant has another APO installation, on which material requirements planning and production planning and detailed scheduling are conducted for the respective plant.

SAP APO supports the following functional areas:

- Cross-corporate interaction on a strategic, tactical and operative planning level
- Collaboration with logistics partners, from order receipt to stock monitoring to product shipping
- Maintenance of relationships with customers as well as business partners
- Continuous optimization and rating of the efficiency of the logistics network

Figure 5.11 shows how the individual production logistics components of SAP APO (Demand Planning, Supply Network Planning as well as Production Planning & Detailed Scheduling) interact. These production logistics components will be presented below in more detail.

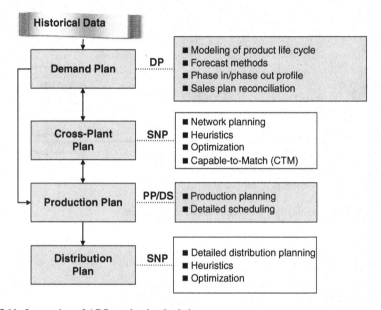

Fig. 5.11 Interaction of APO production logistics

5.4.1   Demand Planning

Demand Planning is a tool to generate forecasts for market demand of a company's products. Several factors are taken into consideration that affect demand in one way or another. The result is the demand plan.

In comparison to demand planning in SAP ERP, the Demand Planning function in SAP APO has been expanded considerably to include the following:

- Expanded user-specific planning layouts and interactive planning books offered for all planning functions.

- You can take internal departments, those of subsidiaries as well as external partners into consideration in the planning and forecasting process (collaborative planning).
- An expanded set of forecasting processes are available that also supports macro functions.
- Planning results are always consistent across all levels.
- The inclusion of historical and planning data is not based in LIS (see Volume II, Chapter 5, "Controlling and Reporting"), but rather on the considerably more efficient SAP NetWeaver Business Warehouse (BW), which is an integrated component of SAP APO and is used in the system for all analytical tasks.
- Forecast models and results can be predefined and subjected to self-defined tests.
- You can consolidate demand plans of various departments using a consensus-based approach.
- Market information and management instructions can be included using promotion or forecast corrections.
- You can phase demand plan products in and out to depict segments of the life cycle.
- Through integration with supply network planning, reconciliation between locations and the available logistics network can be done as early as the demand planning phase.
- Using *Duet* software (with which Microsoft Office can be used with SAP software), Excel integration gives managers direct access to demand planning.

Figure 5.12 shows the basic process of Demand Planning and the interaction with Supply Network Planning. A complete planning cycle includes the following steps:

1. **Preparatory activities**
   Preparation and configuration comprise several steps:

   – Planning report administration
   – Master data configuration
   – Planning book design

   Planning areas are the central data structures in Demand Planning; *planning books* are directly based on a planning area. A planning area contains the following information: the planned quantity unit, planned currency, information on currency conversion, storage time horizon, aggregate levels for data storage, key figures that flow into the planning, and information on aggregation and disaggregation of key figures. Using the master data of Demand Planning, you determine the levels on which demand plans can be generated, edited, aggregated and disaggregated in your company. For the *design* of the planning book, you can determine how the planning images will present the data to demand planning employees or groups, i.e., which characteristics, key figures and data streams are to be displayed, either as a table or graph. In addition, you can define *macros* that are activated when a planning book is opened.

Once preparation and configuration have been carried out, you have laid the foundation for the use of Demand Planning.

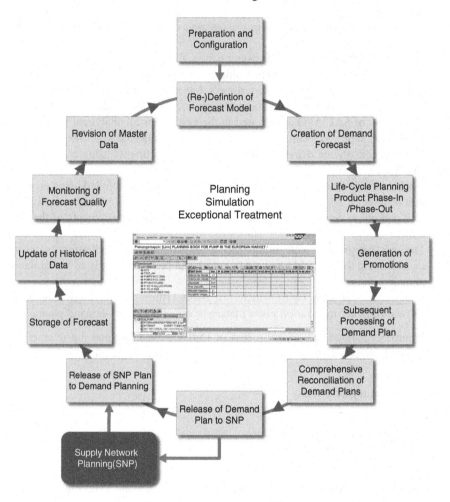

Fig. 5.12 The demand planning cycle

2. **Definition of the forecast model**

   For the definition or redefinition of forecast models, you create a *forecast profile* that controls the automatic calculation of the forecast. The selection of which forecast model will be used depends on several factors, such as whether historical data is available on the product for which a forecast is to be generated, or whether various causal factors for product demand must be taken into account. During definition, you can decide for which products the forecast will be generated and what models will be used.

   The system supports you in selecting the method either by automatically determining the most suitable forecast model or though analysis functions that indicate errors in the test runs using various methods. The settings are then saved as a master forecast profile. Subsequently, you can define an Alert profile that displays *alerts* (warning messages) if forecasting errors occur.

3. **Creation of the demand forecast**
   In this step, you create a new demand forecast or edit an existing one. This procedure can either be triggered interactively by the user or as bulk processing in the background. Demand Planning supports a variety of planning methods: top-down planning, middle-out planning and bottom-up planning. The planning result is consistent at all times and on all levels, since every change is automatically transferred to the other planning levels via aggregation or disaggregation.

4. **Life-cycle planning and product phase-in/phase-out**
   Every product has a life cycle that is characterized by a launch, growth, maturity and discontinuation phase. These phases can be defined in this step. Using special profiles, you can, for example, define that demand for a product will rise slowly to the target value during its phase-in (*phase-in-profile*). When products are being launched in certain regions, you can utilize *like profiles*, which generate a demand forecast based on the demand history of the product in other sales regions.

5. **Generation of promotions**
   In the event that promotional measures (promotions) are undertaken for certain products, you can include their effects in the forecast. Promotions can either be one-time or repeated events. Examples of promotions include Christmas sales, special campaigns, contests or trade-show events.
   By planning promotions separately, the employee for demand planning can compare the figures in various plan versions with or without the promotion.

6. **Seasonal planning**
   Seasonal planning enables time-based aggregation in seasonal periods, that is, you can aggregate data in a defined time period. The definition can vary from one region to another (when it is summer in Europe, it is winter in South America). The data of several periods of a certain interval (such as a month) can be aggregated into a freely definable season, and data from several seasons can be aggregated into a season year.

7. **Error handling and comprehensive reconciliation of demand plans**
   Subsequent processing of demand plans is done through interactive demand planning. By selecting the planning book, you can depict demand forecasts in the desired form and time period, either as a table or graph. Figure 5.13 shows the depiction of a planning book with historical and forecast data as a table and graph.

8. **Reconciliation of demand forecasts**
   If various organizations within a company or even external partners create demand forecasts, they need to be reconciled with each other before you can continue working on the consolidated data. There are several procedures available for this process:

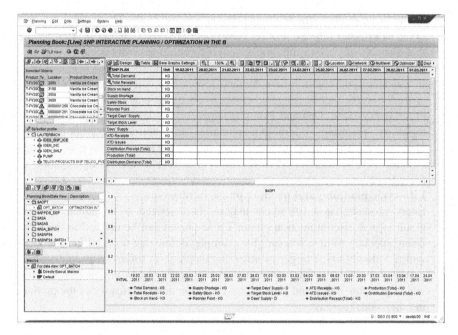

Fig. 5.13 Interactive demand planning

- *Demand Combination* compares forecast key figures of a planning book period by period according to certain criteria and then determines a new key figure that is assumed in the combined forecast key figure.
- The *consensus-based forecast* supports the reconciliation of forecast results from various departments if their forecasts are based on starkly different business goals and time horizons. The system supports the unification process between the different organizations, in order to make it easier to find a consensus for a common demand plan.
- *Collaborative demand planning* is a process that facilitates collaboration between manufacturers and your distributors. Both partners can access the same database and work on a single demand plan. In this way, they can better formulate process flows and profit from a more accurate forecast, batter market transparency, greater stability, reduced stock and better communication.

9. **Interaction with cross-plant planning**
   After releasing a demand plan, a supply chain planner working on cross-plant planning (*Supply Network Planning*, SNP) can access the demand plan data and use it to make decisions regarding supply source determination (*Sourcing*), order fulfillment (*Deployment*) and transport. Furthermore, the demand plan can be transmitted to SAP ERP Demand Management.

Once planning is completed in SNP, the final SNP plan is released back to Demand Planning. This enables the demand planner to perform a comparison of his or her restriction-free demand plan with the restriction-based SNP plan.

10. **Saving the forecast and further steps**
    After the SNP results are reconciled with the DP demand plan, you can save them in the info structures of SAP NetWeaver BW. When doing so, you can also store several key figures for a forecast value in one plan version. If, for instance, you create a demand plan for spring 2012 and want to re-forecast it in January, February and March, this can be done in a plan version without overwriting the old data values. Once the data planning cycle is completed, the historical data of the last cycle is updated for further use.
    During the forecast period, you can monitor the quality of the forecast, in order to be able to make any corrections to the forecast model or select another model. The following analytical functions are available in Demand Planning to perform those tasks:

    – Statistic error analysis
    – Error or reconciliation measures for various forecast models
    – Comparison of planned and actual data
    – Display of earmarked key figures in SAP NetWeaver Business Warehouse

    In a further step, which actually initiates the next forecast cycle, you can adapt the master data to the altered circumstances by, for example, defining new products, product lines, customers or markets, or determining products or lines to be phased out.

5.4.2   Cross-Plant Planning

Cross-plant planning with the APO component Supply Network Planning (SNP) provides planning functions and optimization options that enable you to make tactical planning decisions for the areas of procurement, production, distribution and transport. Using them, you can decide on supply sources or production locations based on a global and consistent model. Unlike Demand Planning, these functions do not have corresponding functions in SAP ERP Production Logistics.

SNP plans the product flow in a logistics network using optimization procedures that employ planning constraints and penalty costs to generate an optimal solution for material procurement and production situations. The results are optimal purchasing, production and distribution decisions, reduced stock, as well as improved customer service when employing *Vendor Managed Inventory* (VMI).

Supply network planning with SNP has the following advantages:

• Cross-plant, medium-term rough operations planning
• Simultaneous planning of procurement, production and distribution
• Planning of critical components and bottleneck resources

- Simultaneous planning of material stock and movement, and finite planning of production, warehouse and transport capacities
- Cross-plant optimization of resource exploitation
- Prioritization of requirements and stock inflow
- Cooperative procurement planning with an Internet connection
- Detailed distribution planning (deployment) with the generation and check of feasible stock transfer
- Grouping of deployment stock transfer into common transport means (Transport Load Building, TLB)

The demand plans represent the starting point for the SNP optimization process. SNP planning is then carried out in two phases (see also Fig. 5.14):

1. The optimization of distribution and demand fulfillment in your own network through suitable transfer of unused requirements between the storage locations or production plants
2. The distribution of available stock to the requirement-generating locations through deployment, in order to cover any open requirement quantities

Supply Network Planning generates a feasible, medium-term plan for the coverage of estimated sales volumes. This plan contains various stock sources:

- Quantities that must be transported between two locations (for example, from the distribution center to the customer or from the production plant to the distribution center) and that lead to a stock transport requisition
- Quantities to be produced that lead to planned orders for Production
- Quantities to be procured that lead to purchase requisitions

Figure 5.14 depicts the detailed process steps in Supply Network Planning. The following steps are involved:

1. **Definition of master data**
   During preparation and configuration, master data is to be defined for the optimization and heuristics processes. Other required master data relates to demand and stock forecasting, safety stock planning, deployment and its optimization, and the Transport Load Builder. Entering master data generally only needs to be done once, and only updated when changes are made to the network structure or processes.
2. **SNP planning run**

   - *Optimization-based planning* searches through all feasible plans to find the one with the least expensive total costs rating. The total costs consist of the following costs: production, procurement, storage and transport, costs for increasing production, storage, transport and handling capacities, costs for an insufficient safety stock quantity or late delivery, and shortfalls.
   - The *heuristic run* determines the procurement requirements for all locations, one after another, and summarizes them. Subsequently, the valid supply sources and quantities to be procured are determined according to

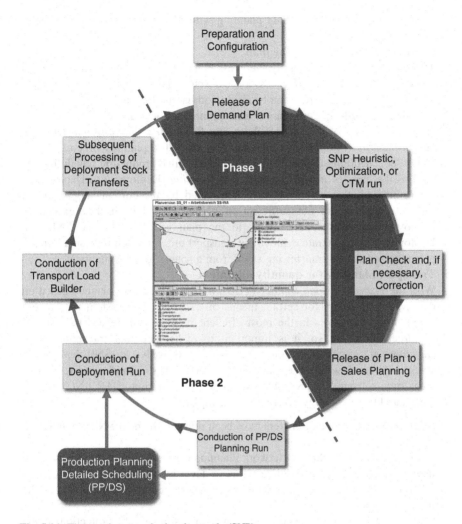

Fig. 5.14 The supply network planning cycle (SNP)

percentages or priorities. However, a plan that has been generated in this way is not necessarily feasible; planning employees must then use the capacity reconciliation function to make it feasible.

– Using *Capable-to-Match-Planning* (CTM), you can conduct a multi-level, finite planning of requirements. Unlike the Optimizer, CTM planning uses a heuristic method, in which, for example, the order of requirements and selection of procurement alternatives is influenced by priorities. Because individual production and distribution levels are not viewed successively, but rather simultaneously, CTM planning guarantees the generation of a timely, feasible plan.

The planning run of cross-plant planning with SNP can be conducted using a variety of procedures:

3. **Correction and release of the plan**

   The plan generated by Supply Network Planning can now be checked by a planning employee and, if necessary, corrected. The plan is then released to Demand Planning, where it can be used to adjust the forecast (see Fig. 5.12).

4. **Detailed scheduling, deployment and quantity allocation**

   At the beginning of the second phase, the *Production Planning and Detailed Scheduling run* (PP/DS, see next section) is started to plan resource allocation and order scheduling in detail. The results of detailed production scheduling are used as the basis of *detailed distribution planning* (Deployment). Deployment determines what requirements can be covered by the current supply. If the supply quantities are not sufficient to cover demand or if they exceed it, Deployment performs adjustments to the plan generated by the SNP run. Deployment can generate deployment transport orders, which represent a transport requirement for transferring stock from a source to a destination.

5. **Transport planning for quantity allocation**

   With the *Transport Load Builder* (TLB), transport loads are planned based on deployment transfers for certain means of transport. The system tries to put together the transports in the most efficient manner with regard to cost and resources. In doing so, it is important to make sure that transport means capacities are exploited as much as possible. Not only can SNP deployment stock transfers flow into the planned transports, but so can planned replenishment orders from SAP Supplier Network Collaboration (SAP SNC), which means additional synergy effects in the company.

After deployment stock transfers have been revised, the next SNP planning run can be started.

A further function in Supply Network Planning is *safety stock planning*. This can be used to reach a predefined delivery service level by maintaining a safety stock for all intermediate and finished products in the respective locations across the entire logistics network.

With the aid of safety stocks, you can assure stock-based protection of the logistics network from external and internal influencing variables, such as:

• Uncertainty of forecast customer demands
• Production disruptions
• Fluctuations in transport times
• Deviations in the planned replenishment time due to supplier problems

Within the context of safety stock planning, two questions need to be addressed:

• Where should the safety stock be kept?
• How high should safety stock be?

There are two methods of planning safety stock. Standard safety stock planning calculates safety stock using empirical values; extended safety stock planning automatically calculates safety stock using the following data, some of which you can specify as mandatory in the SNP system:

- The service level that is to be attained by maintaining the calculated safety stock
- Current demand forecast and historical data regarding the demand forecast
- Safety stock methods and demand types
- Replenishment lead time (RLT) as key figures

The system uses historical data to calculate the forecast error for the demand forecast and replenishment lead time. The actual safety stock calculation can be a relatively complex process in an extensive logistics network, because the number of possible decisions increases exponentially with the number of possible locations.

5.4.3   Production Planning and Detailed Scheduling (PP/DS)

Production Planning and Detailed Scheduling in SAP APO is a tool similar to *Material Requirements Planning* in SAP ERP (see Sect. 5.3.3). The primary areas of application are:

- Generation of procurement recommendations to cover product requirements for in-house production or external procurement
- Determination of resource allocation and detailed planning and optimizing of order dates

Production Planning and Detailed Scheduling is based on several kinds of orders that can come from a variety of sources:

- **Customer orders and stock**
  Customer orders and stock data are transferred from the ERP system to the APO system.
- **APO orders**
  In SAP APO, requirements are automatically generated by the various planning processes (such as SNP) that can be saved as orders. Such APO orders include purchase requisitions, stock transport requisitions and stock transfer reservations (external procurement from an internal supplier), as well as planned orders for in-house production. Execution of planned orders is always done via SAP ERP; the APO orders are transferred back after planning has been completed, where they trigger production orders, purchase orders or order processing.
- **Production orders, purchase orders, project orders, maintenance orders and inspection lots**
  As in the case of customer orders, these orders are transferred from the ERP to the APO system.

Figure 5.15 shows the cycle of Production Planning and Detailed Scheduling. After master data has been defined (such as for models, products and resources in the preparatory and configuration step), the actual planning cycle begins, including

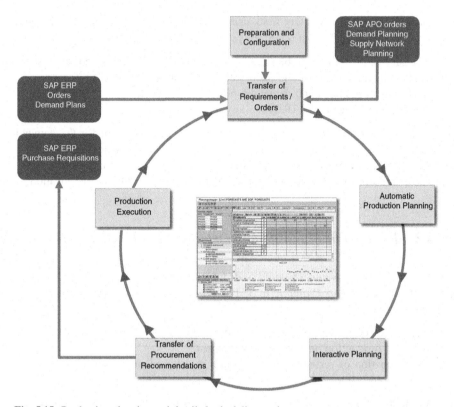

Fig. 5.15 Production planning and detailed scheduling cycle

the transfer of various order and requirements data to Production Planning and Detailed Scheduling. This is generally done through corresponding release steps, such as in Demand Planning or Supply Network Planning (see Fig. 5.14).

In the subsequent production planning run, planning is conducted for selected objects. In the case of a smaller amount of data, this can be done in online mode. However, for larger planning tasks, it is better to use the background planning function. In the production planning run, you can either use heuristics, PP/DS optimization or detailed scheduling functions to find solutions, regardless of the planning task. It is also possible to have different planning tasks performed in several steps in sequence and with various heuristics or functions. For example, you could first perform procurement planning with a product heuristic for products whose requirements are not covered, and then a sequence optimization for the respective bottleneck resources.

Heuristics solve planning problems for selected objects (such as products, resources or orders) using a certain planning algorithm. You can utilize heuristics in interactive planning as well as production planning. The following are examples of where heuristics can be employed in production planning (there are more than 20 processes in total):

- Planning standard batches
- Planning shortage quantities
- Reorder level procedure
- Multi-level planning of individual orders
- Rescheduling bottom-up or top-down procedures

Optimization in Production Planning and Detailed Scheduling enables you to conduct finite planning, resulting in a feasible production plan. For this, stock and requirements as well as dates, resources and activity sequencing are taken into account in a single target function to minimize total costs of the production run. With the optimization function, you can optimize production dates and resource allocation, taking into consideration parameters such as production intervals, setup times and costs, delay costs and mode costs (fixed and variable costs of an activity). In an optimization window, optimization identifies a plan in which the desired result, for instance minimal setup times, is achieved as favorably as possible. To do this, the system may vary the start dates and resource allocation of procedures. Result evaluation is done based on the sum of weighted times and costs that are especially critical to planning. In the course of optimization, the system attempts to reduce the value of the target function, that is, to find a plan in which the various times and costs are as small as possible, corresponding to their weighting. A solution that is optimal from every perspective is generally not possible. For instance, shortening the setup times may lead to an increase in production intervals. In addition, hard and soft ancillary constraints that the system must adhere to exist under certain circumstances during optimization. An example of a hard constraint would be working hours that cannot be exceeded, and a soft constraint would be a confirmed delivery date.

It can generally be said that a longer optimization runtime leads to better planning results and that the necessary minimum runtime increases along with the complexity of the planning problem.

For interactive planning, which includes subsequent processing of the planning results produced by the heuristic or optimizer, there are several interactive tools available, some of which can be integrated in an overall view:

- Product view: Demand/stock situation for a location product
- Product planning table: Overview of the planning situation of several products (see Fig. 5.16)
- Detailed scheduling planning board: Gantt diagrams that show the time position of activities, operations and orders on resources
- Resource planning table: A table depicting resource capacities

The further steps in the product planning process pertain to the transfer of procurement recommendations to the ERP system, in which production orders or purchase requisitions are generated. Integration with production control includes action control, order splitting, production confirmation and planned order management, which are only mentioned here for the sake of completeness, as they are not part of production logistics.

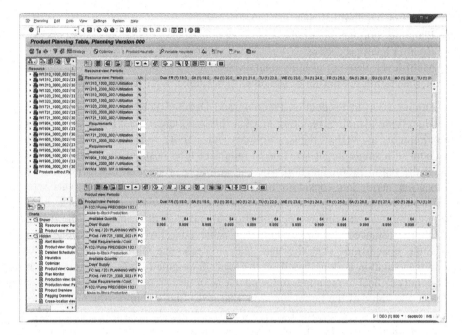

Fig. 5.16 Product planning table of production planning and detailed scheduling in SAP APO

5.5 Summary

This chapter has presented the various tools of production logistics in SAP ERP and SAP APO. For standard processes in the realm of demand planning and production planning, the functions integrated in SAP ERP will certainly suffice. However, if more complex planning procedures or central demand planning from a local system are required, you will need to use SAP APO.

In the next chapter, we will provide an introduction to distribution logistics, which directly depends on the results of optimally functioning production and, from the process view, follows the provision of the finished products.

Chapter 6
Distribution Logistics

Distribution logistics refers to the business process of selling goods, including delivery, transport to the customer and subsequent invoicing. It links production logistics of a company (and/or its external procurement department in the case of externally produced goods) with the demands (orders) of customers. The primary objective of distribution logistics is the efficient provision of goods for customers under set criteria, such as quantity, time and price. The main processes of distribution logistics comprise sales, shipping and invoicing. In addition to the operative tasks, a logistics organization must also fulfill planning-based functions, such as designing optimal distribution networks or selecting the location of a distribution center.

In this chapter, we limit our scope to the sale of goods and the availability of material. The requirement for logistics services in such cases stems from customer orders, whose function we will explain in detail. The sale of externally procured goods is explored in Chap. 4, "Procurement logistics", and sales and demand planning in Chap. 5, "Production Logistics".

Distribution logistics in this chapter focuses on sales processes, ranging from the sales order, to delivery, to the invoicing of traditional sales materials.

> **Professional Services**
> Service processes such as those at consultancies, accounting firms, law offices, temporary employment agencies and IT service providers, belong to the realm of *Professional Services*, which is not within the scope of this book.

6.1 Fundamentals of Distribution Logistics

The first link in the distribution logistics chain is sales. Based on a sales order, goods are either produced, externally procured or taken from the warehouse for delivery (transport) to the customer. Produced or procured goods are also added to the warehouse as stock and thus made accessible for distribution.

J. Kappauf et al., *Logistic Core Operations with SAP*,
DOI 10.1007/978-3-642-18204-4_6, © Springer-Verlag Berlin Heidelberg 2011

6.1.1 Business Significance

Distribution logistics is the integrated planning, design, control, performance and check of the entire material flow and corresponding information flow, beginning with the customer (demand) and usually spanning several production and distribution steps, up to the delivery of the goods to the customer and subsequent invoicing, as illustrated in Fig. 6.1.

Fig. 6.1 Distribution logistics as part of the overall logistics chain

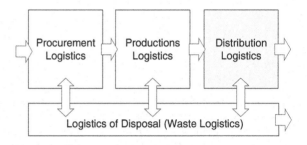

The following explains how you can map the demands of distribution logistics using SAP solutions.

6.1.2 Systems and Applications

As mentioned above, the distribution logistics process generally begins with a sales order. The *sales order*, from a business standpoint, refers to the operative activities surrounding the sale of goods. The following sections will provide an overview of which SAP systems can be involved in sales and distribution processes, what tasks they perform and how their integration contributes to process-oriented distribution. Figure 6.2 shows an example of the interaction between SAP ERP and SAP CRM in the distribution process.

6.1.2.1 SAP ERP

For distribution logistics, the SAP ERP *Sales and Distribution* (SD) component offers comprehensive and convenient functions for the processing, optimization, monitoring and analysis of process-oriented distribution chains. With regard to the distribution process, the functions in SAP ERP can be divided into the elements Sales, Shipping, Invoicing and Customer Complaint Processing (see Fig. 6.3).

SAP ERP offers the following distribution functions:

- Availability check
- Shipping point and route determination
- Transfer of demand to Planning

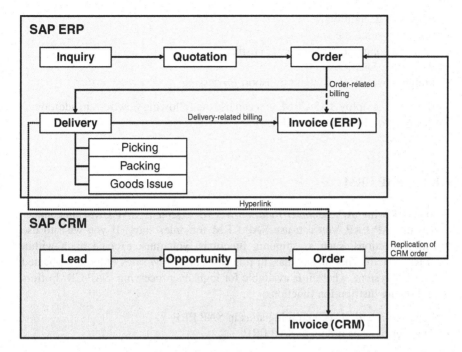

Fig. 6.2 Interaction of SAP ERP and SAP CRM in the distribution process

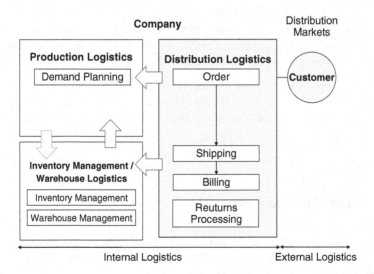

Fig. 6.3 Distribution logistics in SAP ERP

- Integration with production, procurement and warehouse logistics
- Triggering of direct orders (individual orders) from the sales order
- Scheduling of dispatch and transport

- Shipment generation
- Integration into the warehouse
- Commissioning and goods issue posting
- Packaging functions
- Freight cost calculation and transport processing

If you also employ SAP CRM, you can use the following functions in addition to the distribution functions in SAP ERP.

6.1.2.2   SAP CRM

If you work with *SAP Customer Relationship Management* (SAP CRM) in connection with SAP ERP, you can use SAP CRM for order entry. If you wish to use logistics functions, such as shipping functions, you must execute them within the ERP system. This means that, in this case, the CRM sales order is replicated in the ERP system, where it is available for logistics processing. SAP CRM offers the following distribution functions:

- Integration with distribution logistics in SAP ERP
- Transfer of sales orders to SAP ERP

When using integrated sales with SAP CRM and SAP ERP, you can select from the following distribution functions:

- Transfer of the CRM sales orders to SAP ERP for further processing
- E-commerce (Webshop) as an input channel for sales orders in SAP CRM (Internet Sales) and/or SAP ERP (Internet Sales R/3 Edition)

The above refers to the most significant distribution functions specific to SAP ERP and SAP CRM as well as the comprehensive functions. We will now take a closer look at the important functions, to provide a basic understanding of them and their interaction.

6.2   Master Data in Sales

Building on Chap. 3, "Organizational Structures and Master Data", we will present the sales-specific functions of the customer master, material master and customer material information records (see Fig. 6.4).

Figure 6.4 provides an overview of how data from the customer master, customer material info record and material master record are utilized for sales orders. The data belonging to a master record is taken from the corresponding record when a sales order is created by entering the corresponding number, such as a customer number, and stored in the sales order (document).

Fig. 6.4 Master data in the sales process

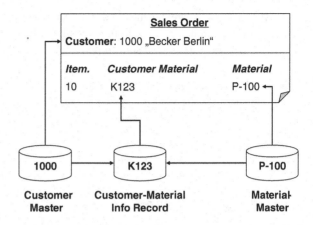

6.2.1   Customer

Information pertaining to individual customers of a company is stored in the customer master records. In addition to the name and address of the customer, the *customer master record* includes information on the proper currency, payment conditions and the names of contacts at the customer's company. Because the customer is also considered a business partner on the debts side of accounting, the customer master record contains accounting data, such as the control account of G/L accounting. Maintenance of the customer master record is thus usually done by the Sales Department as well as Accounting. Detailed information on such general and accounting-specific data can be found in Chap. 3, "Organizational Structures and Master Data".

Depending on business requirements, the data stored in the customer master record may apply only to certain organizational levels. For this reason, the customer master record is composed of four areas that enable individual maintenance of the relevant data: *company code, sales organization, distribution channel* and *division*. The combination of sales organization, distribution channel and division forms the *sales area* within which you can maintain data relevant to sales (sales and distribution data). In the company-code-specific data segments of the customer master, you can enter such accounting-related information as customer bank account data. It is important to note that you can only settle a sales transaction if you have recorded the payer, in the accounting sense, in the system.

The data relevant to sales in your company that can be entered per sales area generally includes the respective customer contact, pricing data, delivery priorities and shipping conditions. Currency and payment conditions valid for a sales transaction are entered into the *sales and distribution data*. Figure 6.5 shows the sales-related distribution areas of a customer.

In addition to the sales-specific data, you can also store *partner functions* in the customer master record (see Fig. 6.6).

Partner functions are business partners that assume certain roles in a sales transaction with a customer. Such functions include the significant and indispensable sales process roles of the sold-to party, ship-to party, payer and bill-to party.

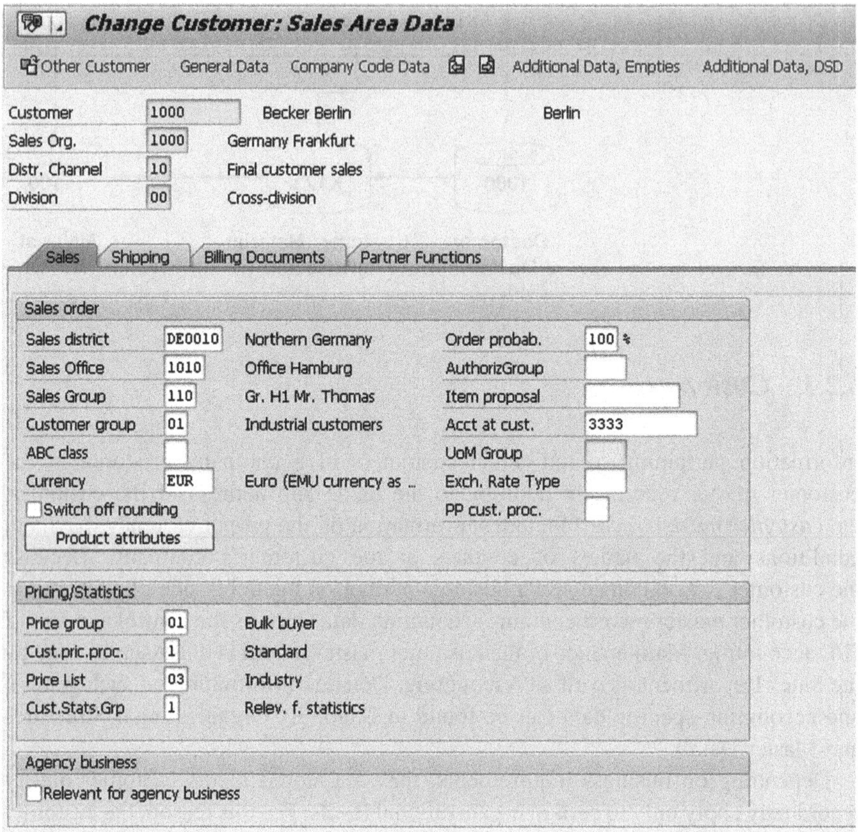

Fig. 6.5 Sales and distribution data of a customer in SAP ERP

Furthermore, you can freely define additional partner functions, such as alternative payers to be determined when posting a customer invoice with reference to a sales document for that customer. Generally, business partners can take on a variety of roles, depending on the business necessity.

During a sales procedure, the customer master record first determines the customer address of a company, then the actual recipient of the goods, followed by the payer, and finally the invoice recipient. In order to be able to use partner roles, the corresponding master record must exist for the respective partner, and the relationships—the partner functions—must be maintained in the respective customer master.

The partner functions roles are assumed in the sales documents as default values, and can be tailored there as needed.

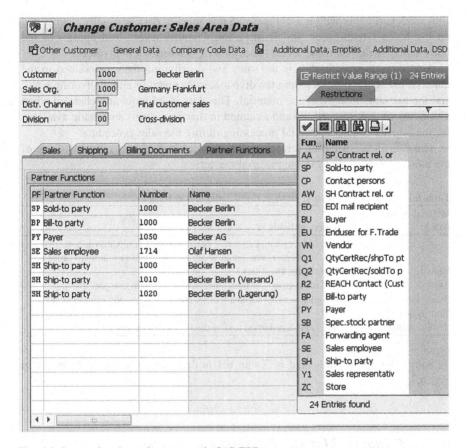

Fig. 6.6 Partner functions of a customer in SAP ERP

6.2.2 Material

The sale of products requires the existence of a *material master record*, which contains the descriptions and control data of all products and components that a company procures, produces, stores or sells. Thus, it is the central source for access to material-specific information. The integration of all material data in a single master record eliminates the problem of redundant data and allows joint use of the stored data, by the Sales Department as well as other corporate areas.

The basic characteristics and views of the material master record were covered in Chap. 3, "Organizational Structures and Master Data". In this section, we provide a brief description of the sales-specific information in the material master that can either be maintained on the level of the *Client* or those of the *Sales Organization* and *Distribution Channel*. Data on the Client level contains information that is equally applicable for every company and organization (such as the Plant and Distribution). Plant data includes information relevant to individual production sites or departments within a company. Sales-specific data is usually maintained on the Sales Organization and Distribution Channel levels, and applies to the documents created on those levels (sales orders, deliveries and invoices).

Among sales-specific data is the sales and distribution data maintained in the material master record, including the division, sales quantity unit, delivering plant, and the tax classification of the material. This data is subsequently checked when a sales document is generated and assumed in that document, thus made available for the logistics and commercial processing during the sales procedure.

The data and distribution screens and fields cited in Table 6.1 represent some of the most important views in the material master.

Table 6.1 Sales and distribution views in the material master record

Material master view (tabs)	Important fields
Sales 1	Units of measure, delivering plant, material group, division, tax classification and quantity agreements such as minimum order and delivery quantities
Sales 2	Statistical groups, product hierarchies and item category group
Sales/plant data	Weights, settings for availability check, batches and shipping data
Text in sales and distribution	Sales and distribution text in the respective language

Some of the data indicated in Table 6.1 is depicted as a system example in Fig. 6.7.

6.2.3   Customer-Material Information Record

The customer-material information record serves as an information source for Sales and Distribution, and represents individual customer data pertaining to a material. The data contained in the customer-material info records has a higher priority than data from the customer and material master, and is used for document processing (see Fig. 6.8).

In addition to the fields displayed in Fig. 6.8, the customer-material info record contains data such as the delivering plant, delivery priority, minimum order quantity and further information on partial delivery options and sales document control on the item level. This data is shown in Fig. 6.9.

You can use the data of the customer-material info record for such tasks as the creation of a material item in a sales document. The system then uses the customer material number you have entered to determine the internal material number under which the material is maintained in the sales unit. With the aid of the text search function, you can also find text stored in the customer-material info record and use it in the sales procedure, for instance to use certain expressions or provide a distribution employee with supplemental text.

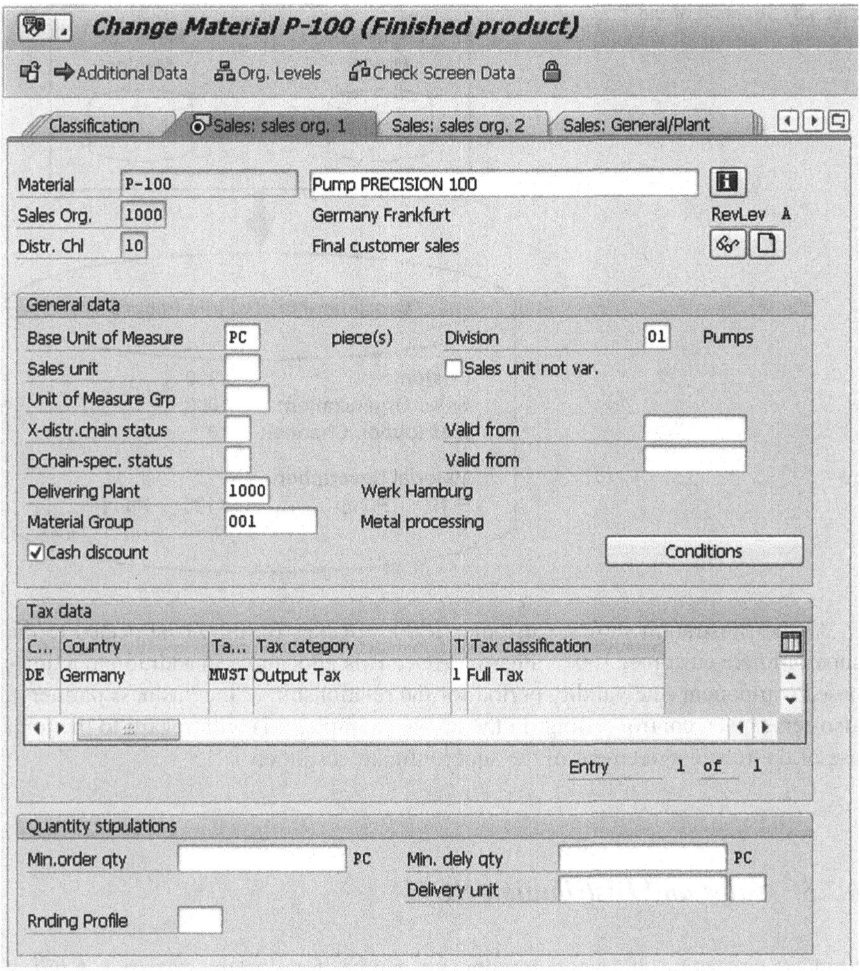

Fig. 6.7 Data for "Sales Organization 1" regarding a material in SAP ERP

6.2.4   Customer Hierarchies

Customer hierarchies provide the structured mapping of relationships, for instance within an affiliated group or purchasing association. Customer hierarchies can be used in sales processing for business partner determination and pricing. For example, a superordinate customer such as an umbrella organization can be consulted in order to determine conditions (prices). The data can also be evaluated for rebate determination, if the customer or superordinate association (hierarchy) has a rebate agreement in place. Figure 6.10 shows the primary elements of the customer hierarchy.

Fig. 6.8 Structure of the
customer-material info record

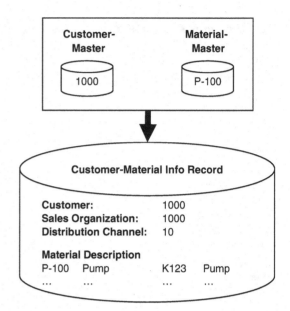

As the illustration shows, Customer 1000, "Becker Berlin", is allocated to the superordinate customer 1007, "Pharma AG". This allocation, in addition to a time-based component (the validity period for the relationship of the business partners), also serves as a control setting; in the above example, it is with regard to the joint use of the rebate agreement of the superordinate customer.

6.2.5   Sales and Distribution BOM

Products consisting of several components can be stored in the system as a *bill of materials (BOM)* (see Fig. 6.11). In the bill of materials, you can maintain the components (that is, the material), such as for a sales product. You can also save documents such as explosion diagrams or sales-related images in the BOM.

The BOMs have a validity and revision level, via which you can maintain and execute planned changes to the BOM structure. For instance, if you want to use the BOM for a sales order, you can save it under Usage 5, ("Sales and Distribution"), and thus have created a sales and distribution BOM. In the standard version of SAP, you are provided with eight BOM usages that you can adapt to your needs, or you can also define new ones.

When you enter the BOM in a sales order, a *BOM explosion* takes place and the components are entered into the sales order as a subordinate item (see Fig. 6.12). *Pricing* can either be done on the main item or the subordinate item, that is, the components. If pricing is done on the components, the sales price is determined from the sum of the individual prices for the components. For *transfer of requirements*

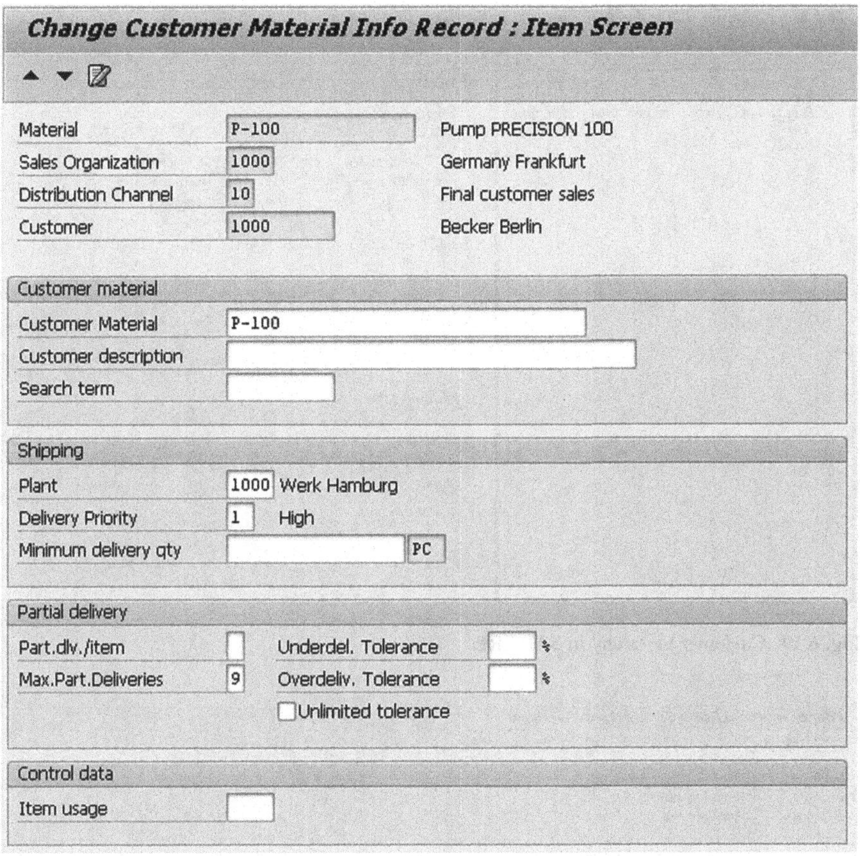

Change Customer Material Info Record : Item Screen

▲ ▼ 🖉

Material	P-100	Pump PRECISION 100	
Sales Organization	1000	Germany Frankfurt	
Distribution Channel	10	Final customer sales	
Customer	1000	Becker Berlin	

Customer material

Customer Material	P-100
Customer description	
Search term	

Shipping

Plant	1000	Werk Hamburg
Delivery Priority	1	High
Minimum delivery qty		PC

Partial delivery

Part.dlv./item		Underdel. Tolerance		%
Max.Part.Deliveries	9	Overdeliv. Tolerance		%
		☐ Unlimited tolerance		

Control data

Item usage	

Fig. 6.9 Customer-material info record in SAP ERP

to Material Requirements Planning, the same applies: Transfer of requirements is done either on the main item or subitem. You can control via the item types the pricing, transfer of requirements is based on the settings of the schedule line. For a further illustration of the document structure (header, item data, schedule line data), see Fig. 6.14.

6.3 Sales

The *sales document* structures the sales process, since you can define the significance of the various documents in the system configuration. The main types of sales documents are inquiries, quotations, orders, agreements (contracts and scheduling agreements) and complaints (free-of-charge delivery, credit and debit memo requests and returns). Based on the sales document, you can create and edit

Maintain Customer Hierarchy, Standard Hierarchy, Date: 17.02.2011

Cust. hierarchy	Custo...	Loc	Sales area
▾ 🔲 Pharma AG	1007	DE-60311 Frankfurt	1000/10/00
· 🔲 Becker Berlin	1000	DE-13467 Berlin	1000/10/00

Assignment

Higher-level customer

Cust.		
Sales organization	1000	Germany Frankfurt
DistrChannel	10	Final customer sales
Divis.	00	Cross-division

Customer

Cust.		
Sales organization	1000	Germany Frankfurt
DistrChannel	10	Final customer sales
Divis.	00	Cross-division

| From | 17.02.2011 | Valid to | 31.12.9999 |

✔ Transfer

Application log

T... | Message Text | Cust. | Sal

Fig. 6.10 Customer hierarchy in SAP ERP

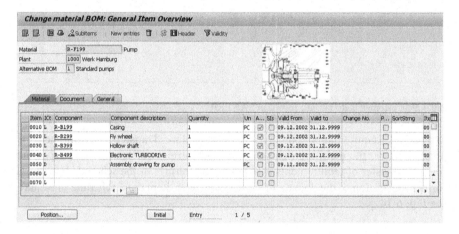

Change material BOM: General Item Overview

Material	R-F199 Pump
Plant	1000 Werk Hamburg
Alternative BOM	1 Standard pumps

Material | Document | General

Item	ICt	Component	Component description	Quantity	Un	A...	SIs	Valid From	Valid to	Change No.	P...	SortStrng	Its
0010	L	R-B199	Casing	1	PC	☑	☐	09.12.2002	31.12.9999		☐		00
0020	L	R-B299	Fly wheel	1	PC	☑	☐	09.12.2002	31.12.9999		☐		00
0030	L	R-B399	Hollow shaft	1	PC	☑	☐	09.12.2002	31.12.9999		☐		00
0040	L	R-B499	Electronic TURBODRIVE	1	PC	☑	☐	09.12.2002	31.12.9999		☐		00
0050	D		Assembly drawing for pump	1	PC	☐	☐	09.12.2002	31.12.9999		☐		00
0060	L					☐	☐				☐		
0070	L					☐	☐				☐		

| Position... | | Initial | Entry | 1 / 5 |

Fig. 6.11 BOM

follow-on documents, such as delivery and billing documents. In the system configuration, you can fine-tune follow-up actions for sales documents. For instance, for orders of the types *cash sale* and *rush order*, the system generates delivery and billing documents without user interaction as soon as you save the document. An example of the relationships between the individual sales documents can be seen in Fig. 6.13.

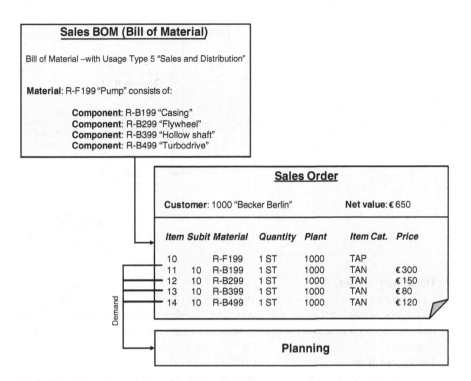

Fig. 6.12 Usage of the sales and distribution BOM in a sales order

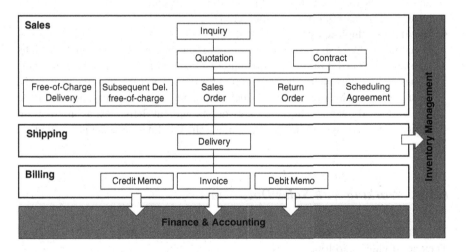

Fig. 6.13 Overview of the sales procedure

Figure 6.14 illustrates the document structure of the sales documents, which we will examine in more detail below. The general structure of the documents is presented in Chap. 2, "SAP Business Suite". A document consists of the following elements:

Fig. 6.14 Document
structure

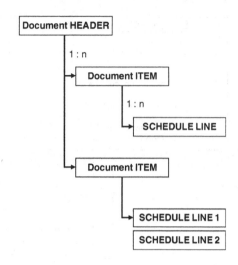

- **Document header**

  You can enter general information in the header (header data) that applies to the entire document, including the items and schedule lines. For instance, such information might include the partner (customer), document currency and pricing procedure.

- **Document item**

  In the document item (item), you can enter data that applies only to that specific item, for instance the material number and, if required, any ship-to party or payer that deviates from the header. Furthermore, you can enter the plant, storage location, and material price on the item level.

- **Document schedule line**

  A document item can consist of one or more schedule lines. Additional schedule lines are generated if the desired quantity of a material is not available on the date requested by a customer. All data required for further processing of the delivery is entered on the schedule line level. Such data includes the quantity, delivery data and confirmed quantity. In the case of orders without deliveries, as in credit or debit memos, the system does not generate schedule lines, as none is required.

6.3.1   Working with Sales Documents

The individual sales procedures are structured in the system using *sales documents*. They are divided into four groups:

- Inquiries and quotations
- Orders
- Agreements, including quantity and value contracts and scheduling agreements
- Complaints, such as free-of-charge deliveries and credit memo requests

These sales documents will be examined in more detail below. To make it easier to use the system, you can create follow-on documents, such as deliveries, directly from a sales document. The sales document type, such as a standard order, is used to decide what follow-on documents are possible or necessary. An example would be a rush order, which automatically generates a delivery when the order is saved in the system.

The following basic functions are available in order processing (see Fig. 6.15):

- The availability check function checks the material quantity ordered by your customer to see if it is available and can thus be confirmed.
- For the transfer of requirements, your Planning Department is informed of material quantities and dates on which the order is to be delivered.
- Delivery scheduling uses the desired delivery date of the customer to check whether the goods can be delivered on time.
- Based on the settings that you have made in the system, the customer-specific prices are determined using the condition master data and added to the sales document.
- The credit limit check provides information on the credit rating of a customer.
- After the sales procedure is completed, you will generally create documents such as an order confirmation.

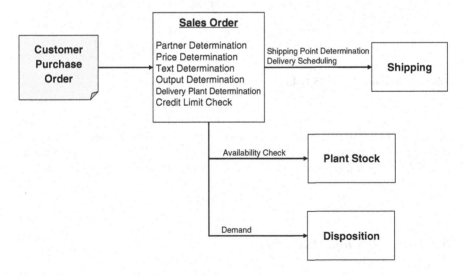

Fig. 6.15 Basic functions in the sales order

In the system configuration, you can set which functions are to run automatically or manually (with user interaction).

The created sales documents are independent documents, but can also have a reference to each other. In other words, they can represent parts of a document chain. For instance, a customer can make an inquiry to your company that you enter in the system as a document. Based on this inquiry, you can subsequently enter the offer desired by the customer. You can then convert it into an order if your customer accepts the offer. The system supports you by copying the data from the appropriate document, in this case the quotation, to the follow-on document, the sales order. The goods included in the order are delivered and invoiced. After delivery of the goods, the customer complains of a technical defect, and a delivery free of charge is also generated based on the sales order. As you can see, the system allows you easy interaction without the need to painstakingly enter the data again.

The illustrated document chain, spanning the inquiry, quotation, sales order, delivery, invoice and free-of-charge delivery, forms the *document flow* (see Fig. 6.16).

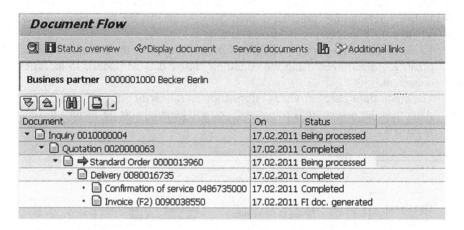

Fig. 6.16 Document flow in SAP ERP

Using this document flow, you have the opportunity to find and navigate documents associated with other documents at any time. This is useful when your customer has a question about his invoice but only has the order number at hand. Based on that number, you can use the document flow to find the corresponding invoice.

6.3.2 *Inquiry and Quotation*

In the pre-selling phase, you can enter the business events of the inquiry (from the customer) and the quotation (issued to the customer) in the system and use them for evaluation purposes and to create follow-on documents, such as an order based on a quotation.

Figure 6.17 shows an example of an inquiry. Your customer contacts you to find out about a certain product. For instance, he might want to know if the product is in stock and at what price it can be purchased. You create an inquiry for this event, to record the data and forward the requested information to the customer. If your customer subsequently asks you to submit a quotation, you can use the data from the inquiry as a basis, and create the quotation with reference to the inquiry (see Fig. 6.18).

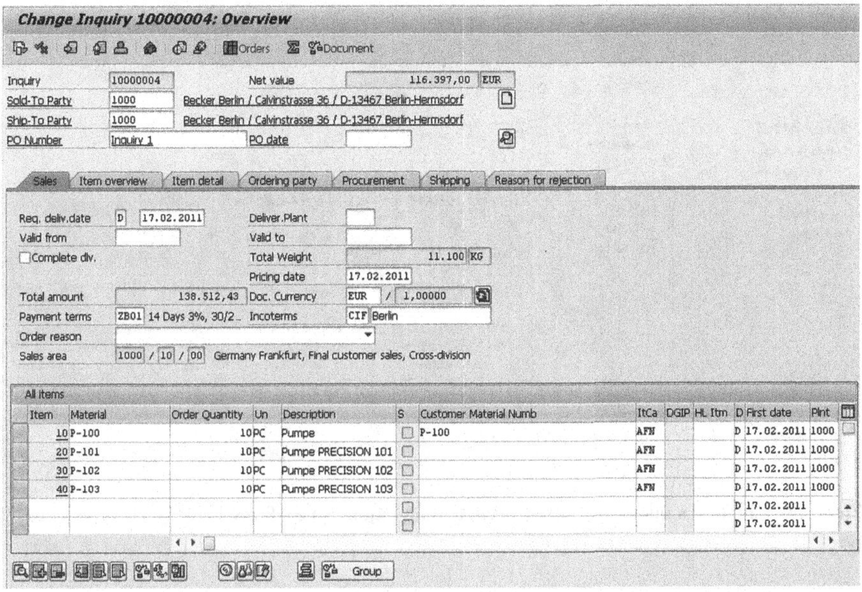

Change Inquiry 10000004: Overview

Inquiry	10000004	Net value	116.397,00	EUR
Sold-To Party	1000	Becker Berlin / Calvinstrasse 36 / D-13467 Berlin-Hermsdorf		
Ship-To Party	1000	Becker Berlin / Calvinstrasse 36 / D-13467 Berlin-Hermsdorf		
PO Number	Inquiry 1	PO date		

Sales | Item overview | Item detail | Ordering party | Procurement | Shipping | Reason for rejection

Req. deliv.date	D 17.02.2011	Deliver.Plant		
Valid from		Valid to		
Complete dlv.		Total Weight	11.100	KG
		Pricing date	17.02.2011	
Total amount	138.512,43	Doc. Currency	EUR / 1,00000	
Payment terms	ZB01 14 Days 3%, 30/2..	Incoterms	CIF Berlin	
Order reason				
Sales area	1000 / 10 / 00	Germany Frankfurt, Final customer sales, Cross-division		

All items

Item	Material	Order Quantity	Un	Description	S	Customer Material Numb	ItCa	DGIP	HL Itm	D	First date	Plnt
10	P-100	10	PC	Pumpe		P-100	AFN			D	17.02.2011	1000
20	P-101	10	PC	Pumpe PRECISION 101			AFN			D	17.02.2011	1000
30	P-102	10	PC	Pumpe PRECISION 102			AFN			D	17.02.2011	1000
40	P-103	10	PC	Pumpe PRECISION 103			AFN			D	17.02.2011	1000
										D	17.02.2011	
										D	17.02.2011	

Group

Fig. 6.17 Overview of an inquiry in SAP ERP

Creating an Inquiry
Figure 6.17 shows the inquiry of Customer 1000. The company Becker Berlin inquires about Customer Material P-100 (spiral pump). Based on the entry of the customer material, the system checks the customer-material info record to find Material P-100, which your company carries, and enters it in the document. The functions of the sales document are performed on the basis of Material P-100.

Since the structures of the sales and distribution documents are the same with regard to data technology, that is, they only differ by the type and corresponding functions, you can copy data from a previous document (in this case the inquiry) into the follow-on document (here, the quotation). You can decide whether you want to transfer all of the data or, via item selection, only parts thereof into the follow-on document. This is recommended if, for instance, your customer has inquired about complementary products, but now only wishes to have a quotation for one particular product. The same applies for quantities about which your customer has inquired and for which a quotation is to be issued.

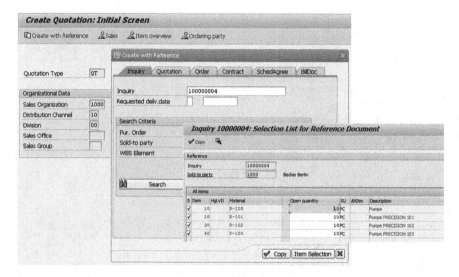

Fig. 6.18 Creating a quotation with reference to an inquiry in SAP ERP

Creating a Quotation with Reference
Figure 6.18 illustrates how you can create a quotation based on an inquiry. If you know the inquiry number, you can enter it directly. Otherwise, there are several search criteria available to aid you in finding the inquiry. You also have the possibility of either assuming all of the data or individual items of the inquiry via item selection to use in the quotation.

The result of this procedure is the creation of a quotation (see Fig. 6.19), which you can edit if necessary before transferring it to the customer in electronic or paper form.

The generated quotation then represents a legally valid offer to your customer that obligates your company to supply the goods indicated at the price offered. You have already entered the data necessary for further processing, such as the business partners (customer, ship-to party, bill-to party and payer), products and quantities, prices and schedule lines (the point in time in which the product is available for shipping).

If the customer places an order, the data contained in the quotation can, in turn, be used as the basis for the generation of the sales order. This means that, as demonstrated in Fig. 6.18, you can create the sales order with reference to the quotation. The same functions for generating a document with reference are also available here.

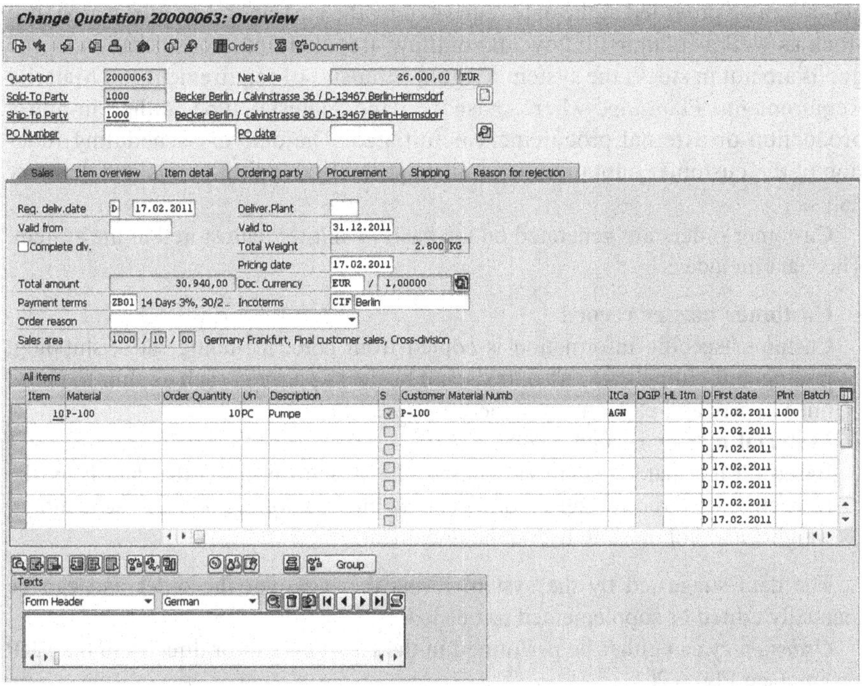

Fig. 6.19 Overview of a quotation in SAP ERP

Editing a Quotation
Figure 6.19 displays the generated quotation in edit mode. In this case, you have created Quotation 20000063 based on Inquiry 10000004, and before sending the final quotation to the customer, you wish to check the data again and make any necessary changes.

6.3.3 Order Processing

The *sales order* is a contractual agreement between your company and your customer for the delivery of goods or services at stipulated prices, quantities and dates. The system performs various functions within the context of order processing. For example, pricing is executed to determine the sales price. The price can be determined on a customer-specific basis, for instance based on a contract, or from a price list (see Fig. 6.23). The availability of goods is also determined, meaning the system automatically checks whether the ordered goods can be provided at the date requested by the customer. For this, warehouse stock as well as planned inflow and outflow are taken into consideration. If the goods are not in stock, the system triggers a transfer of requirements to Material Requirements Planning, where, based on the material type, either in-house production or external procurement is initiated. The delivery scheduling function of the customer quotation determines possible shipping points and delivery dates.

Customer orders are generated on the basis of master data stored in the system. This data includes:

- **Customer master record**
  Customer-specific information is copied from here, including sales, shipping, pricing and billing data. Also, texts and business partners such as ship-to party, bill-to party and payer are assumed from it.
- **Material master record**
  Based on the material number entered, the system retrieves the data from the respective material master record, such as information on delivery scheduling, weights and volumes, to use in the document.

The data suggested by the system forms the basis for the order and can be manually edited or supplemented as needed.

Order entry can either be performed in the CRM system or directly in the ERP system (see Fig. 6.20).

Fig. 6.20 Sales and distribution process in SAP CRM and/or SAP ERP

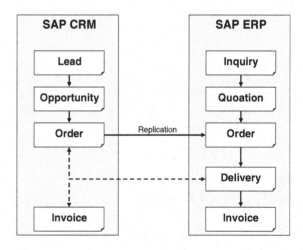

When you open the sales order entry function in SAP CRM, the sales order is replicated in the ERP system for logistics processing. The deliveries generated in SAP ERP are shown in the CRM system within the context of document flow, and you can also access them from the CRM system. Figure 6.21 shows order entry in SAP CRM, and Fig. 6.22 shows order entry in SAP ERP.

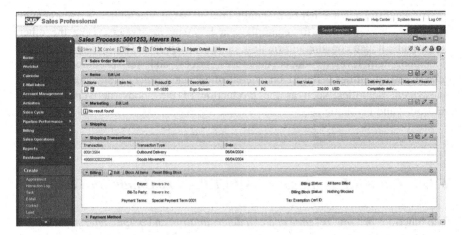

Fig. 6.21 Sales order in SAP CRM

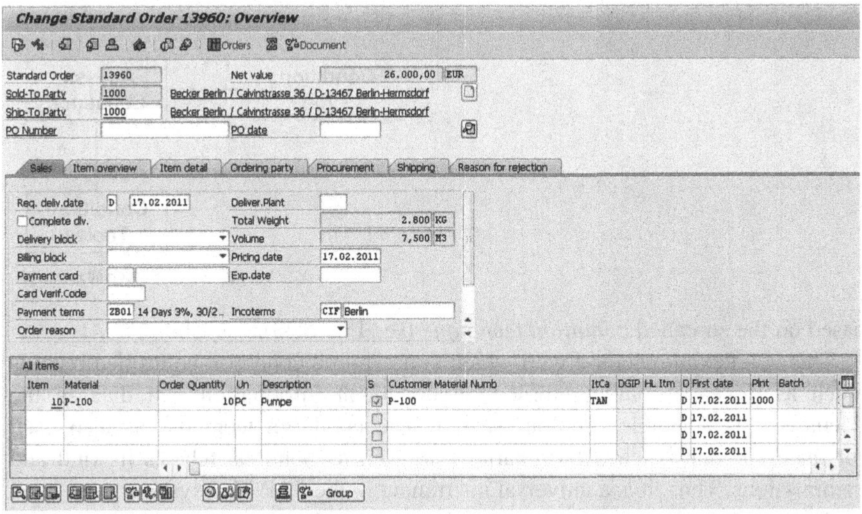

Fig. 6.22 Sales order overview in SAP ERP

Editing a Sales Order

Figure 6.22 shows Sales Order 13960 for the customer, the company Becker Berlin. This is a standard order with a sales order item for Material P-100 having a quantity of 10 units. The calculated total net value of the order amounts to €26,000. In the above case, the ship-to party 1000 is also the company Becker Berlin. Of course, you have the option of entering an alternative ship-to party. This can be done by entering another customer number or by manually entering the alternate delivery address.

The sales order can either be manually created based on a reference document or electronically generated (for example, via EDI) by the system. Like the other documents, it also contains information on the header level (business partner), the document item level (products, quantities, prices) and the schedule line level (confirmed quantities). Based on a confirmed sales order, the system initiates subsequent activities, such as the production of the goods, delivery, picking, transport, and invoicing.

6.3.4   Central Functions in SAP ERP – The Condition Technique

The functions *pricing*, *revenue account determination* and *output determination* are

Fig. 6.23 Function of the condition technique

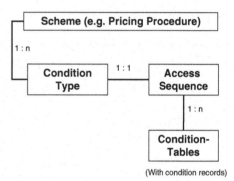

(With condition records)

based on the so-called *condition technique* (see Fig. 6.23).

This is a flexible method with which you can find condition-controlled records (such as prices or outputs) based on document information. All data in the documents is available for your search. The condition technique is also used for material and batch determinations, as well as bonuses, rebates in kind and commissions. Thus, it is a universal instrument in the SAP ERP system.

The condition technique is used to execute the following:

1. **Define a procedure**
   In a procedure, you can define the condition types that you wish to use for determining the condition master records. A procedure can contain several condition types.
2. **Define condition types**
   Condition types can be price elements (such as prices, surcharges, deductions, freight or taxes) or output conditions (such as an order confirmation, delivery note or invoice). An access sequence can be allocated to condition types, which control the conditions determined.
3. **Define condition tables**
   The condition tables are assigned to the access sequence. In the condition tables, the fields are defined that are relevant to determining condition types. The

communication structures (data segments) are filled using the document data, and with them, the condition tables are accessed.

4. **Define access sequences**

The access sequence defines in what order and under what conditions access to condition tables will take place.

The condition technique, as portrayed in the above general description, is utilized in several areas of the SAP system.

6.3.4.1 Pricing

For pricing, you determine the pricing procedure and sequence in the calculation type that you wish to use for pricing purposes (see Fig. 6.24).

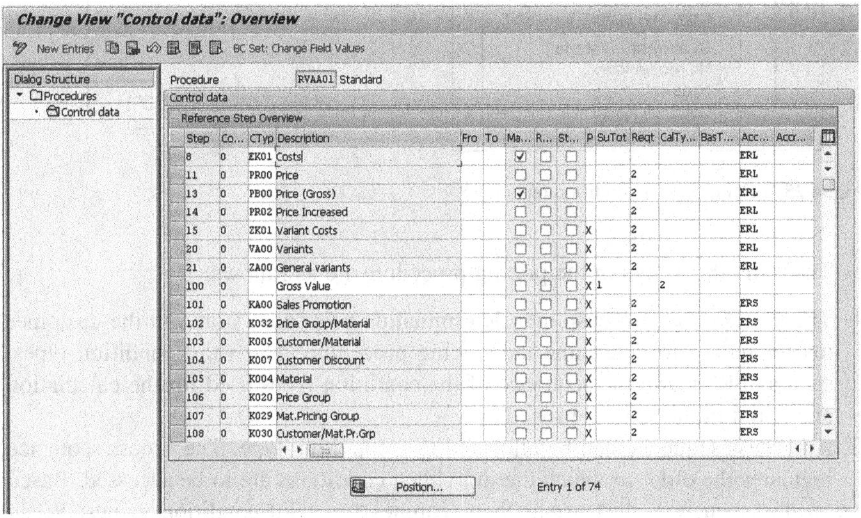

Fig. 6.24 Pricing Procedure in SAP ERP

The pricing procedure is allocated to a sales organization, distribution channel, division, document pricing procedure and customer determination procedure. This allows you to define how the pricing procedure for various business organizations, business document types and customers is determined. For instance you could price quotations based on a different pricing procedure. During price calculation the pricing procedure is used to determine the prices. Using the information from the document (the *communication structures*), the system tries to find a corresponding condition (price, surcharge or deduction, tax) from the condition tables for every condition type via the assigned access sequence. Once a price has been determined, it is utilized in the pricing procedure, as depicted in Fig. 6.26. The pricing function is illustrated in Fig. 6.25.

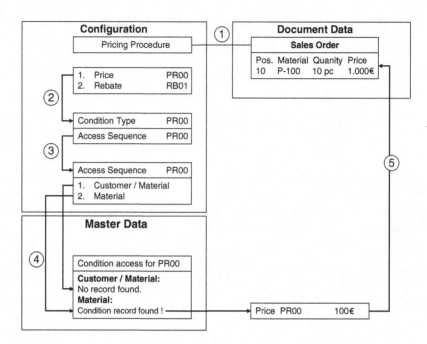

Fig. 6.25 Pricing with a condition technique

The individual steps in the pricing procedure are as follows:

1. The system uses the customer determination procedure stored in the customer master record to determine the pricing procedure. The valid condition types, such as PR00, and the sequence of the condition types used for the calculation are stored in the pricing procedure.
2. An access sequence can be allocated to a condition type. The access sequence regulates the order in which the individual conditions are to be accessed. Based on this sequence, the system then searches for valid condition values. When doing so, it proceeds from specific (such as a customer- specific price) to general information (such as the material price). You can define as many access sequences as desired.
3. In the example in Fig. 6.25, the first access (Customer/Material) is not successful. With the second access, the system then attempts to determine a price.
4. The values stored in the condition record, in our example €100, are then transferred to the sales order, multiplied by the number of the ordered quantity, and put into the condition value.
5. The result is a calculated sales price for PR00 of €1,000. A percent-based or an absolute surcharge or discount can be manually added to or determined for the calculated price. On the last calculation level, the system determines the valid tax rate.

In Fig. 6.26, we see the result of a pricing procedure on the item level of the sales order for Material P-100 with the quantity of 10 units. The system has determined the corresponding pricing procedure according to the above, retrieved a condition record from the master data via the access sequences and added the information to the document. Document pricing also provides an analysis with which you can track why a price, surcharge or discount was determined for the calculation. The determined values thus offer complete transparency.

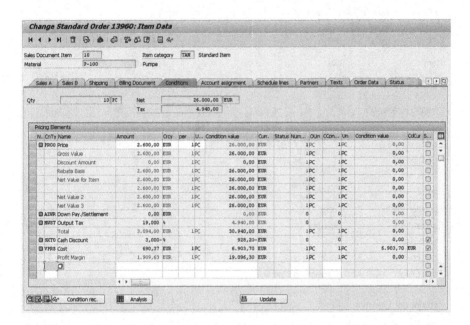

Fig. 6.26 Price overview for a sales order in SAP ERP

The output procedure is allocated to the respective document type and is thus determined when a document is created (such as a sales order). It contains the conditions (in this case, the message output type) that may be used for document processing. Using the conditions, you can fine-tune the settings, for example by only allowing a document to be printed if it has been completed. You can formulate individual conditions to adapt the system behavior to your needs.

6.3.4.2 Revenue Account Determination

For *revenue account determination*, you can set the accounts in which billing prices, surcharges and discounts will be assigned. Revenue account determination also utilizes the condition technique. In the account determination procedure, you can set the criteria for account determination, and allocate the account determination procedure as a follow-up to the billing document type. In the access sequence, you can set the order in which the fields from the billing document are to be consulted for account determination. You can also fine-tune the accounts for a variety of business transactions. You can assign an account key (such as revenue, sales deduction, freight revenue, sales deduction rebate, value-added tax or reserves) to the respective prices, surcharges and discounts. These settings are consulted when the billing document is generated to determine the corresponding account (see Fig. 6.27).

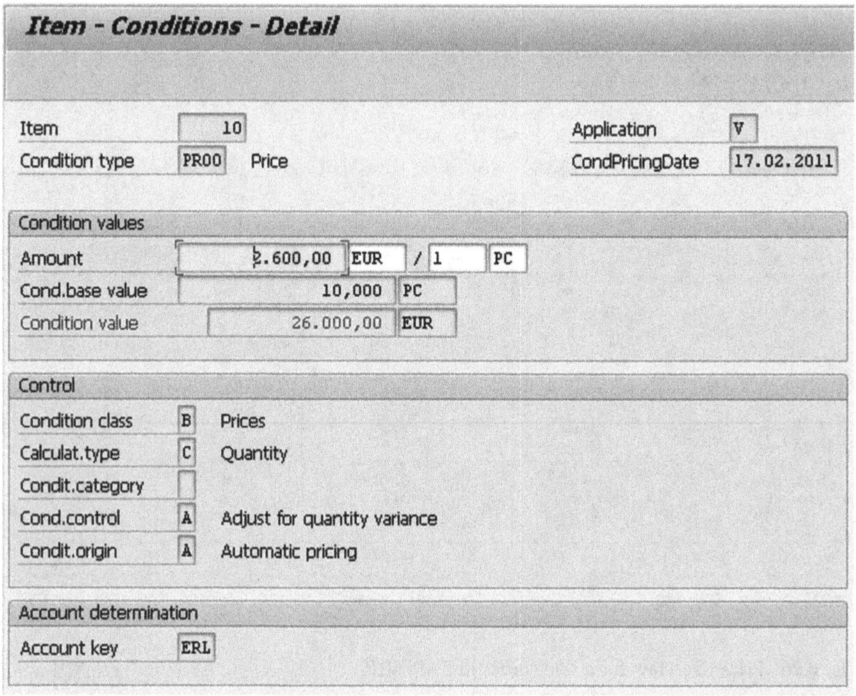

Item - Conditions - Detail

| Item | 10 | | Application | V |
| Condition type | PROO | Price | CondPricingDate | 17.02.2011 |

Condition values

Amount	2.600,00	EUR	/ 1	PC
Cond.base value	10,000	PC
Condition value	26.000,00	EUR

Control

Condition class	B	Prices
Calculat.type	C	Quantity
Condit.category		
Cond.control	A	Adjust for quantity variance
Condit.origin	A	Automatic pricing

Account determination

| Account key | ERL |

Fig. 6.27 Revenue account in the billing document

6.3.4.3   Output Determination

Output determination is a component in SAP ERP that serves a central function (Fig. 6.28). Like pricing, output determination is based on the condition technique. The purpose of output determination is to define a valid output record based on the document information (such as BA00 order confirmation). In the output determination, you can define the output type and the communication medium (printout, fax, telex, EDI, e-mail, workflow, ALE; see Fig. 6.29). Special forms of communication include special event forms that activate their own programs. For instance, the output determination can also be used to trigger interfaces. In addition, printer, output language and recipient of the output are defined (see Fig. 6.29). This data is taken from the condition master records and can be overridden manually in the document. In Fig. 6.28, you can see an example of an output determination procedure.

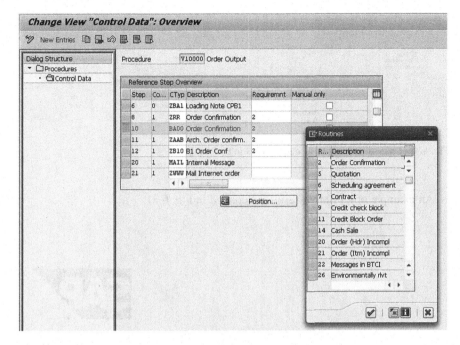

Fig. 6.28 Output determination procedure in SAP ERP

You can assign a print form and print program (see Fig. 6.30) to the output condition type . This allocation ensures that the document data is forwarded to the corresponding print program for printing and transferred to the print form for formatting.

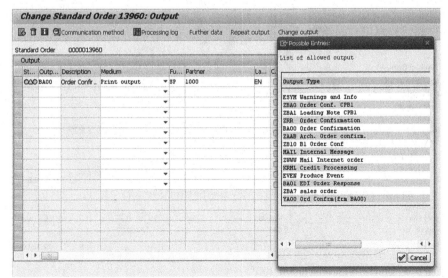

Fig. 6.29 Output determination in a sales document in SAP ERP

Change View "Processing routines": Overview

New Entries Form

Dialog Structure
- Output Types
  - Mail title and texts
  - Processing routines
  - Partner functions

Output Type BA00 Order Confirmation
Application V1 Sales

Processing routines

Medium	Program	FORM routine	Form	PDF/SmartForm Form	Type
Print output	SD_SDOC_PRINT01	ENTRY		SD_SDOC_FORM01	PDF
Fax	SD_SDOC_PRINT01	ENTRY		SD_SDOC_FORM01	PDF
External send	SD_SDOC_PRINT01	ENTRY	SD_SLS_MAIL_BODY	SD_SDOC_FORM01	PDF
EDI	RSNASTED	EDI_PROCESSING			
Distribution (ALE)	RSNASTED	ALE_PROCESSING			

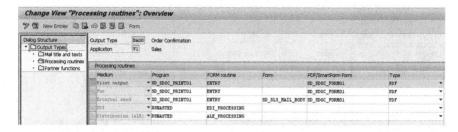

Fig. 6.30 Processing routines of output types in SAP ERP

An example of a printout result is depicted in Fig. 6.31.

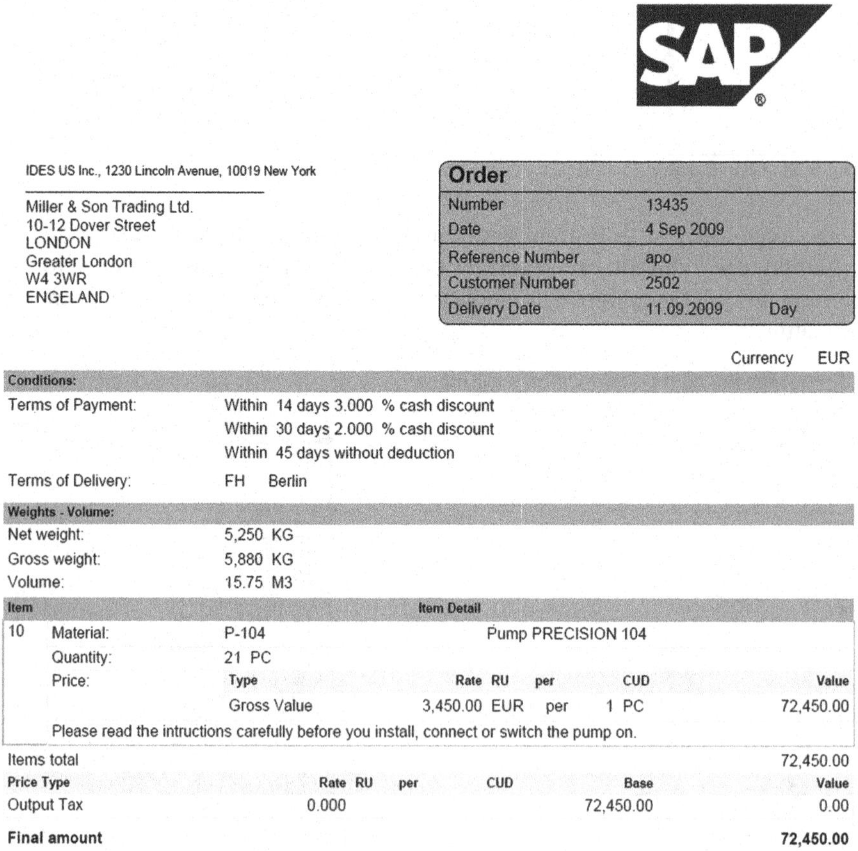

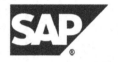

IDES US Inc., 1230 Lincoln Avenue, 10019 New York

Miller & Son Trading Ltd.
10-12 Dover Street
LONDON
Greater London
W4 3WR
ENGELAND

Order		
Number	13435	
Date	4 Sep 2009	
Reference Number	apo	
Customer Number	2502	
Delivery Date	11.09.2009	Day

Currency EUR

Conditions:

Terms of Payment: Within 14 days 3.000 % cash discount
Within 30 days 2.000 % cash discount
Within 45 days without deduction

Terms of Delivery: FH Berlin

Weights - Volume:

Net weight: 5,250 KG
Gross weight: 5,880 KG
Volume: 15.75 M3

Item **Item Detail**

10 Material: P-104 Pump PRECISION 104
Quantity: 21 PC
Price:

Type	Rate	RU	per	CUD	Value
Gross Value	3,450.00	EUR	per	1 PC	72,450.00

Please read the intructions carefully before you install, connect or switch the pump on.

Items total 72,450.00

Price Type	Rate	RU	per	CUD	Base	Value
Output Tax	0.000				72,450.00	0.00

Final amount **72,450.00**

Fig. 6.31 Order confirmation in SAP ERP

6.3.5   Credit and Risk Check

The components for *Credit and Risk Management* (SD-BF-CM) support you in minimizing bad debt loss by assigning customers a credit limit (see Fig. 6.32). You can set the ERP system such that it checks the credit limit stored for a customer in the event of a sales order. If the credit limit has been exceeded, you can block certain procedures, such as the generation of an order confirmation. In addition, your customer's respective employee receives a corresponding notice. A further result of a negative credit check could be the triggering of a workflow to your respective credit employee, who will then check the sales document and can effect a release for it. You can also employ Credit and Risk Management in distributed affiliates, that is, where Financial Accounting is central and the distribution companies are decentral, with each of their own systems mapped.

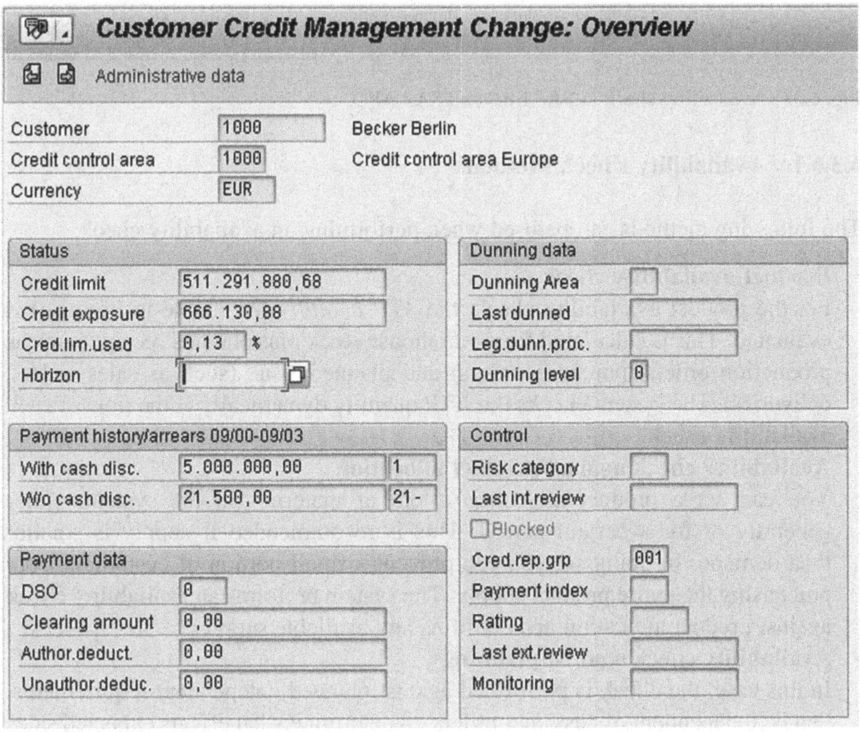

Fig. 6.32 Credit management in SAP ERP

6.3.6   Availability Check (SAP ERP/SAP APO)

SAP generally supports two procedures (see Fig. 6.33) for an *availability check*. The first is an availability check in SAP ERP, the second is one in SAP APO. The major difference between the procedures is that in the first, the inventory in a *local*

ERP system is checked. In contrast, the availability check in SAP APO offers the opportunity to include the global inventory of your affiliates (gATP, *Global Available-to-Promise*).

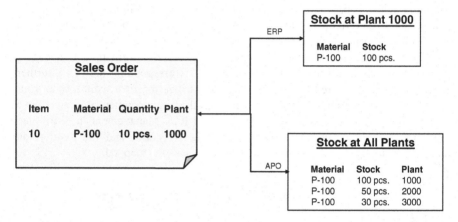

Fig. 6.33 Availability check in SAP ERP und SAP APO

6.3.6.1   Availability Check Methods

The following methods can be used when performing an availability check:

- **Product availability check**
  For the product availability check, the *ATP quantity* (Available-to-Promise) is evaluated. This is calculated from warehouse stock plus planned receipt (such as production orders, purchase orders) and planned issue (such as sales orders, deliveries). The system checks the ATP quantity dynamically at the time of each availability check.
- **Availability check against product allocation**
  You can set a product allocation quota of materials in the system, either generally or for a certain period. This is recommended if supply is smaller than demand. In doing so, you can prevent a small portion of customers from purchasing the entire product supply. The system performs an availability check against product allocation and confirms any available supply.
- **Availability check against planning**
  In this case, the check is performed against planned independent requirements, that is, independent of customer orders, to determine what effects expected sales quantities will have. This type of availability check is used in planning situations.

These methods are available in SAP ERP as well as SAP APO.

When an order is placed, the availability check verifies whether the ordered material can be made available in the quantity desired and at the time requested by the customer. For the availability check, you can define the scope for the individual application areas such that different types of stock, including safety stock, stock in transfer, subcontractor stock and inspection stock, as well as blocked and free stock, are taken into consideration. In addition, you can include elements from inward and outward movement, purchase orders, purchase requisitions, dependent requirements, reserves and sales requirements in the scope of the availability check.

If the availability check indicates that the ordered goods will be available on the requested date, this fact is confirmed in the document (the sales order).

If the result of the availability check is negative, that is, if the ordered goods will not be available on the requested date, a transfer of requirements is triggered if the respective goods are internally produced. This will appear in MRP as a customer requirement, where you can immediately convert it into a planned/production order.

If the products are to be procured externally, you can see the respective mation from the Planning view and trigger a purchase requisition (PR) from there. It will then be turned into a sales order by Purchasing. Alternatively, in special cases, you can trigger a purchase requisition directly from a sales order. This special type of sales order processing is an *individual purchase order*, and is controlled via the schedule line category.

Figure 6.34 shows an availability overview from the Planning view.

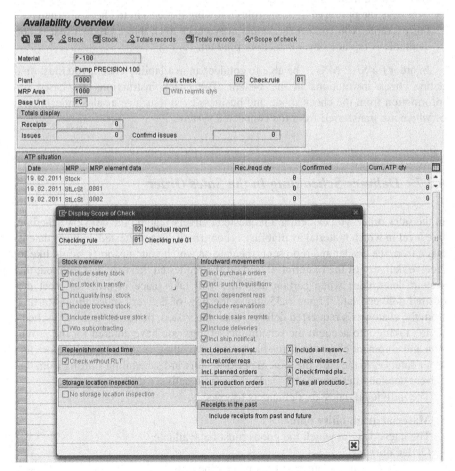

Fig. 6.34 Availability overview in SAP ERP

In addition to the standard methods of availability checking described above, the global availability check in SAP APO supports the following procedures:

- **Combination of methods**
  You can initially conduct a check against product allocation and then let the confirmed quantities be checked again against the ATP quantities.
- **Production**
  Check against production (CTP, *Capable-to-Promise*) assumes that you use the component *Production Planning and Detailed Scheduling* (PP/DS) in the SCM system. In that case, when the order is received, you check it against the PP/DS. You can control this check such that, if insufficient availability results, immediate production or an immediate purchase order is triggered.
- **Multi-level ATP check**
  This type of availability check can be used if you manufacture assemblies and final assembly is done at a later time upon receipt of the sales order.
- **Rules-based availability check**
  You have the option of defining rules allowing the system to check alternatives. For example, you can set the availability check such that a product that is not available is replaced by a substitute product.

In order for SAP APO to be able to conduct an availability check, you must first define check instructions in the system. The check instructions are based on information from the check mode and business event (such as a sales order), both of which are transferred from the connected system.

6.3.7   Delivery Scheduling in the Sales Order

In the sales order, you can enter a requested delivery date on the delivery schedule line level in which material availability is confirmed. The requested delivery date is taken from the header and represents the date on which your customer would like to receive delivery of the material. Based on the requested delivery date entered, the system determines when certain activities will take place in order to meet the desired delivery date. Figure 6.35 shows the flow of delivery scheduling. Starting from the customer's requested delivery date, the system performs backward scheduling , taking into account the scheduling settings you have entered for the provision and shipment of goods. If the backward scheduling results in a date before the order date, i.e. in the past, the system performs forward scheduling starting from the current date and determines the earliest possible delivery date.

The individual dates are defined as follows:

- **Material availability date**
  The latest date on which the goods must be available
- **Order date**
  The date on which your customer placed the order

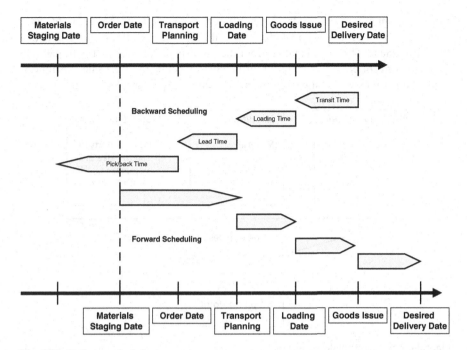

Fig. 6.35 Delivery scheduling

- **Transportation planning date**
  The date as of which transportation activities must begin
- **Loading date**
  The date on which the goods must be ready for loading
- **Goods issue date**
  The date on which the goods have to leave your company for delivery to the customer
- **Requested delivery date**
  The delivery date requested by the customer

You can individually adjust the times cited in Fig. 6.35 with the Customizing feature of the ERP system according to your business needs.

6.3.8 Shipping Point and Route Determination in the Sales Order

After you have indicated the sold-to party and material in the sales order, the system performs a shipping point and route determination on the item level. The shipping point represents the organizational logistics unit responsible for the execution of shipping activities in your company. If a sales order has various items serviced by different shipping points, this leads to a *delivery schedule split* in the subsequent delivery generation, since a delivery is always only processed by one shipping point. The shipping point is determined based on the shipping condition of your customer, the loading group of the material (such as forklift loading) and the delivering plant. This data is taken from the corresponding customer and material master records or customer-material info record and entered in the sales document as a default value, where it can be manually edited if necessary (see Fig. 6.36).

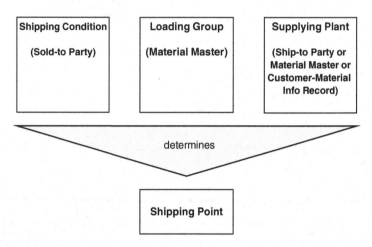

Fig. 6.36 Shipping point determination

For route determination, the system determines an appropriate route on the item level of the sales order based on the departure zone, shipping point, the shipping condition of your sold-to party, and the transportation group of the material and the transportation zone of your ship-to party (see Fig. 6.37). The departure zone is the zone from which the shipping activities are controlled. The receiving zone is the location of the ship-to party. Zones can be freely defined. For instance, they can be organized according to regions (South Zone, North Zone) or postal codes and used for route determination.

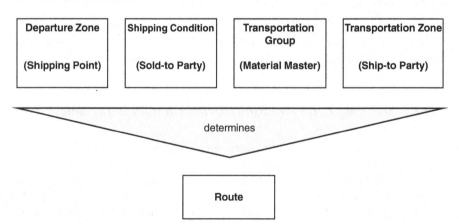

Fig. 6.37 Route determination

Making changes to the sales order, such as to the shipping condition, leads to a new route determination, since you may then no longer be able to use the planned express route, needing instead a more economic route. As you can see, route determination not only has a logistics side, but also a commercial aspect. You can find out more about route determination in Volume II, Chapter 2, "Transport Logistics".

6.3.9 Transfer of Requirements

A *transfer of requirements* (see Fig. 6.38) informs Materials and Requirements Planning of the quantities required in Sales and Distribution to service the sales order. In the case of a material shortage (if the requested quantity of a material is not available), a transfer of requirements can be used in MRP to create planned or production orders for in-house products or purchase requisitions/purchase orders for externally procured products . With regard to the transfer of requirements for sales orders, we differentiate between a *transfer of requirements with individual requirements* and a *transfer of requirements with summarized requirements*. The type of transfer of requirements selected is determined in the respective material master record.

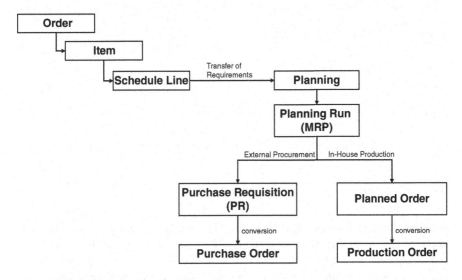

Fig. 6.38 Flow of the transfer of requirements

Information on the Transfer of Requirements
A detailed description of the transfer of requirements for external procurement can be found in Chap. 4, "Procurement Logistics", and in Chap. 5, "Production Logistics" for in-house production.

6.3.10   Backorder Processing

Using *backorder processing*, you can list and manually edit sales documents relevant to requirements. As such, you can allocate quantities from other orders to non-confirmed sales orders, for instance in order to be able to process orders with a higher priority. Backorder processing is only possible for materials identified as

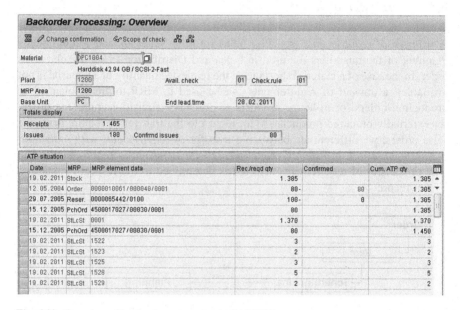

Fig. 6.39 Overview of backorder processing in SAP ERP

individual requirements. Figure 6.39 provides an example of backorder processing.

6.4   Shipping

Shipping is the organizational logistics unit in which physical goods movement is executed. Shipping thus represents the link between an order and the warehouse, and comprises all process steps (see Fig. 6.40) from the picking process, to packaging, to the printing of necessary shipping documents, to goods issue.

In generating deliveries, shipping represents the physical completion of warehouse activities. Once a goods issue is booked, the goods leave your company. Ensuring customer service via the timely delivery of goods is a primary objective of shipping.

6.4.1   Delivery Processing

Delivery (see Fig. 6.41), like the sales order, is a document with a header and item data. When the delivery is created, which you can either trigger individually or in a collective run, data from the sales order is assumed in the delivery document. The delivery itself triggers shipping activities: transfer to the warehouse for picking,

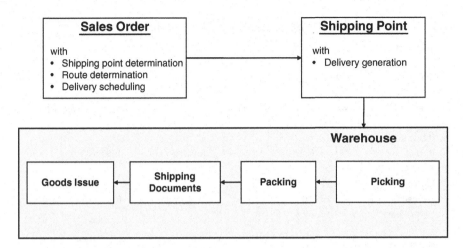

Fig. 6.40 The shipping process

packaging and goods issue. If you use transportation processing in SAP ERP (LE-TRA), deliveries are summarized in shipments. (More information on transportation processing with SAP can be found in Volume II, Chapter 2, "Transport Logistics".)

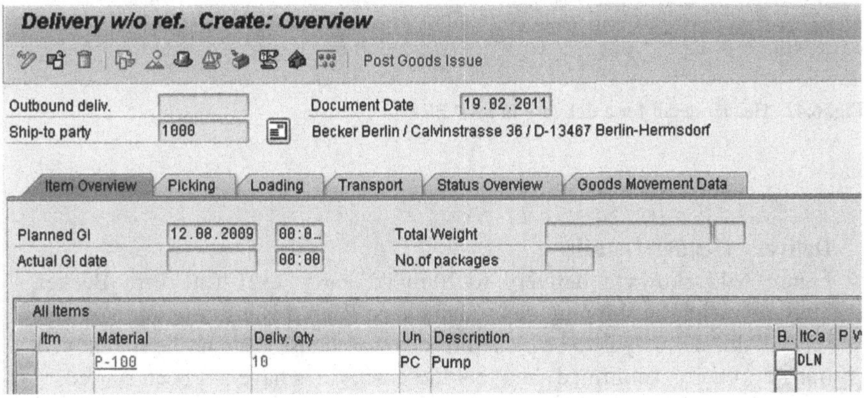

Fig. 6.41 Delivery in SAP ERP

Changing a Delivery
Figure 6.41 shows a delivery to Ship-To Party 1000 (the firm Becker Berlin) with an item, Material P-100, to be delivered to Plant 1000, at a picking quantity of 10 units.

Delivery and the associated shipping activities are generated and processed via a shipping point, the organizational unit of shipping. The shipping point depends on the delivering plant, shipping type and required loading equipment.

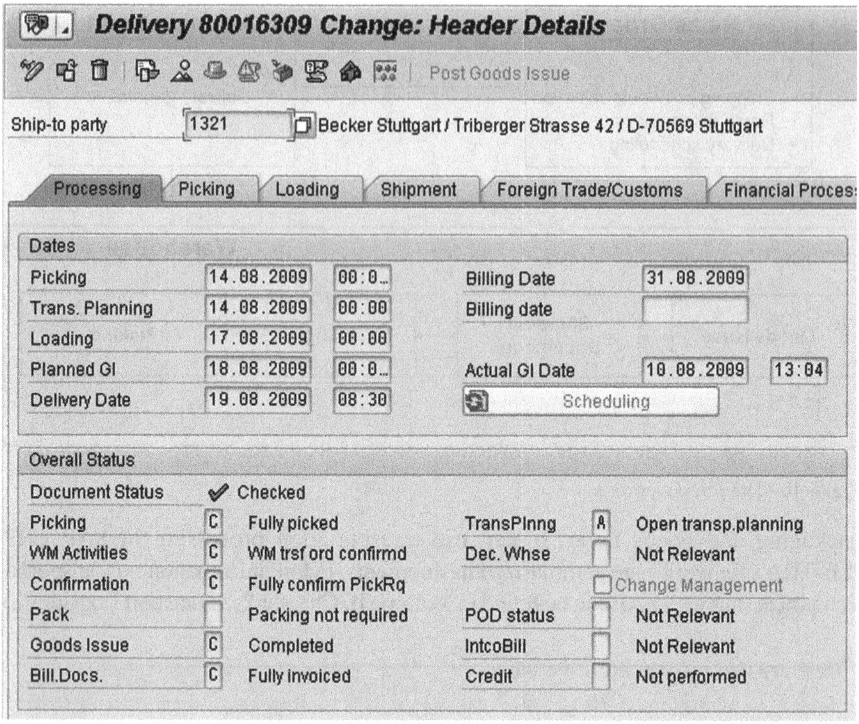

Fig. 6.42 Header detail for a delivery in SAP ERP

Delivery Header Detail
Figure 6.42 shows a delivery to Ship-To Party 1321 (the firm Becker Stuttgart) with the shipping dates and status. With this information, you can see that this delivery is to be picked from a warehouse system (the status WM transfer order is confirmed) and that these activities have not been started.

Deliveries are always created in the ERP system. Generally, you can create a delivery with reference to a sales order (schedule line of the sales order), a stock transfer order (transfer of goods within your company), a subcontractor, a project or without reference. In the customer master, you can define whether your customer allows partial or complete deliveries and/or order combinations. You have a reference to the respective delivery in the connected ERP system from the CRM system, which you can use to create the order (see Fig. 6.43).

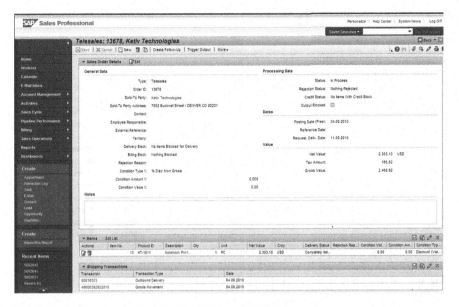

Fig. 6.43 Delivery integration of SAP CRM and SAP ERP

Delivery Integration of SAP CRM and SAP ERP
Figure 6.43 shows the CRM sales order and corresponding ERP delivery,
which was generated based on the replicated sales order in the ERP system.
You can display the ERP delivery directly from the CRM system. The
delivery data is stored in SAP ERP.

You are now familiar with the documents of a delivery. In the following, we will
examine the most important functions within delivery processing.

6.4.1.1 Route Determination in the Delivery

In the section "Shipping point and route determination in the Sales Order" above,
route determination in the case of a customer order has already been examined. At
the time in which the delivery is created, you can execute another route determination, either to confirm the routes suggested from the sales order or to add new routes
due to new circumstances. Figure 6.44 shows the settings for a route determination
of a delivery with variables such as the shipping point, transportation group and
weight group.

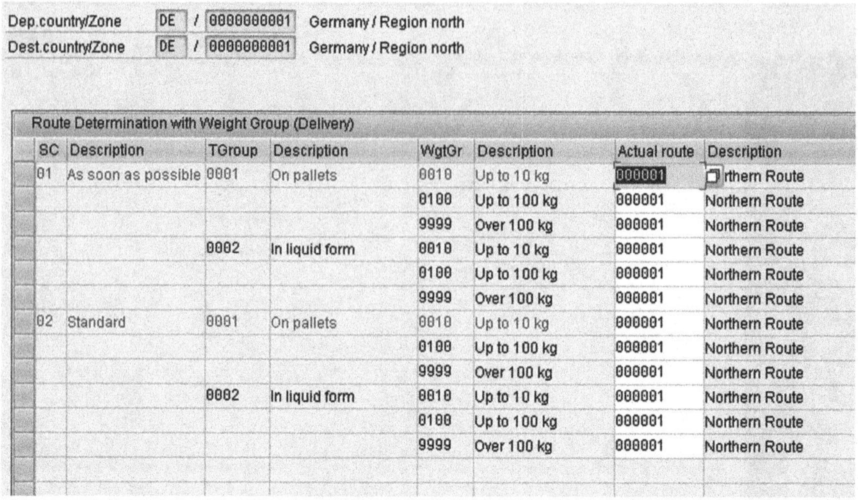

| Dep.country/Zone | DE | / | 0000000001 | Germany / Region north |
| Dest.country/Zone | DE | / | 0000000001 | Germany / Region north |

Route Determination with Weight Group (Delivery)							
SC	Description	TGroup	Description	WgtGr	Description	Actual route	Description
01	As soon as possible	0001	On pallets	0010	Up to 10 kg	000001	⬜ rthern Route
				0100	Up to 100 kg	000001	Northern Route
				9999	Over 100 kg	000001	Northern Route
		0002	In liquid form	0010	Up to 10 kg	000001	Northern Route
				0100	Up to 100 kg	000001	Northern Route
				9999	Over 100 kg	000001	Northern Route
02	Standard	0001	On pallets	0010	Up to 10 kg	000001	Northern Route
				0100	Up to 100 kg	000001	Northern Route
				9999	Over 100 kg	000001	Northern Route
		0002	In liquid form	0010	Up to 10 kg	000001	Northern Route
				0100	Up to 100 kg	000001	Northern Route
				9999	Over 100 kg	000001	Northern Route

Fig. 6.44 Route determination for a delivery in SAP ERP

In addition, you can also employ a *route schedule* (see Fig. 6.45). This is a schedule that enables you to regularly control repeated deliveries from a shipping point to one or more ship-to parties in a certain order along a fixed route. A route schedule generally contains a route, a date and time, the ship-to party and itinerary (optional).

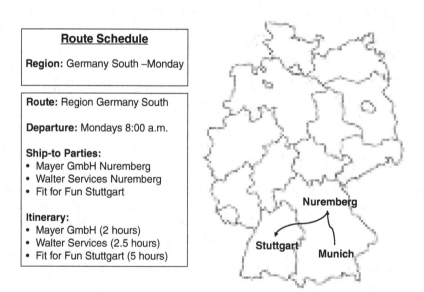

Fig. 6.45 Route schedule

6.4.1.2 Availability Check in Delivery Generation

As described above under "Availability check (SAP ERP/SAP APO)", the basic function of the availability check in delivery processing is not much different from the check performed upon receipt of an order. The only difference is that, at the time the delivery is generated, the material situation is checked against the picking date. An availability check for delivery generation is recommended to determine whether the material situation has changed since the availability check was made for the sales order. If the stock situation has changed since then, you will be notified at the latest when the delivery is generated and you can react accordingly.

6.4.1.3 Picking

When a delivery is created, the confirmed quantity is taken from the sales order schedule line and added to the delivery item as the delivery quantity. This quantity serves as the specification for picking, which is done in a warehouse system. You can transfer the delivery data to the warehouse system where physical picking takes place, that is, where the goods are taken from the storage location according to the picking request, and the picked quantity is reported back to the delivery in the system, which leads to an update of the picking quantity and the picking status.

The *picking status* in the delivery item provides you with a continuous picture of the status of picking activities. WM (SAP ERP), SAP EWM (SAP SCM) or a third-party program can be integrated as a warehouse system. If you use a SAP warehouse management system, you automatically have full integration between the delivery and the warehouse, meaning the current warehouse activity status flows into the delivery. Data from the delivery is transferred via the warehouse interface to the connected warehouse management system, where it is available for picking procedures.

> **Information on Warehouse Management**
> You can find out more about warehouse management in Volume II, Chapter 3, "Warehouse Logistics and Inventory Management".

6.4.1.4 Packing

For packing, you assign a packaging material to the delivery item you wish to pack in the ERP system. The packing dialog can also be accessed using an *RF device* (a radio frequency device, such as a scanner), so that packing functions can also be utilized in the warehouse without requiring access to the system. The packing functions also support you in box packaging (such as a delivery item in a cardboard box, placed on a pallet, which you then load onto a truck for transport). Since you can keep packaging materials (Handling Units, or HUs) as stock, you can also monitor the stock situation of packaging materials (in your own warehouse as well as at customer or carrier locations). In addition, you have the opportunity to trace which packaging material was used in the delivery of a particular material, since the HUs have distinct numbers. Figure 6.46 shows the packing dialog in a delivery.

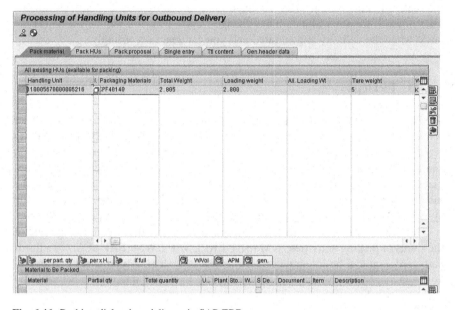

Fig. 6.46 Packing dialog in a delivery in SAP ERP

Packing Procedure in Shipping
Figure 6.46 shows the packing dialog in Shipping. In the above example, Material P-100 (pump) at a quantity of 10 units has been assumed from the delivery to the packing dialog. In this packing dialogue, you mark the material to be packaged and press the 🖫 button (Packing). This leads to a recommendation of suitable shipping materials: In the example above, Shipping Material CPF40140 (pallet), having Handling Unit 110005670000005216 as a distinct identification, has been recommended.

The system can recommend which shipping materials should be used for packing. You also have the possibility of manually overriding the shipping materials recommendation or manually selecting a shipping material.

6.4.1.5 Goods Issue

Goods issue completes the shipping process, which means that the goods leave your company and stock is reduced by the quantities recorded in Goods Issue. Figure 6.47 depicts the flow of the goods issue process.

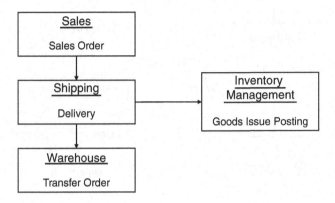

Fig. 6.47 Goods issue process for delivery

As you can see in the above example, based on the sales order, a delivery is created. This delivery is the central element in logistics performance, and is integrated in warehouse and inventory management. You can either post a goods issue manually or collectively, that is, for several deliveries at once. Data from the delivery and the goods issue posting is assumed in the material document of inventory management. You can only alter this data in the delivery itself, and not in the material document, which ensures that the data is current in inventory management. Furthermore, in addition to quantities, goods issue will update any changes in value in the balance sheet accounts. With the posting of a goods issue, the delivery is due for invoicing. In addition to the order or delivery-based goods issue posting, you have further options for goods issue posting within inventory management. They are described in detail in Volume II, Chapter 3, "Warehouse Logistics and Inventory Management".

6.4.1.6 Dangerous Goods Processing

For the transport of dangerous goods, there are legal regulations that you must observe according to the type of goods involved. We refer here to transfer in the broadest sense, including interim storage of dangerous goods in your warehouse, insofar as that warehouse is classified as a dangerous goods warehouse. Dangerous goods processing in the SAP System is done with the application *SAP Environment, Health, and Safety Management* (SAP EHS Management). Figure 6.48 shows the flow of dangerous goods processing.

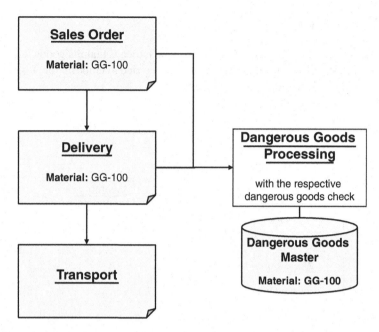

Fig. 6.48 Dangerous goods processing

In the material master, you can set whether a product is categorized as dangerous goods. If this is the case, you allocate the material to the *dangerous goods master*, where the legal *dangerous goods regulations* are stored. Checking of the dangerous goods regulations is either done manually in the document or when the respective document is saved (in our example, the delivery document). Based on the material in the delivery, the triggering of a dangerous goods check identifies that a material falls under the dangerous goods category, and the respective regulations are checked.

6.4.1.7   Shipping Documents

In this section, we will introduce the most common shipping documents. Shipping documents are notices (such as printouts) that have been generated based on the condition technique. The delivery type and item are allocated to an output determination procedure that contains the individual delivery messages. Based on these settings, you can create delivery documents (delivery notes, packing lists) and freight documents (freight lists). The *packing list* is the list that your warehouse uses to package the corresponding materials in the goods issue zone. The *freight list* is used to combine several shipments that are to be loaded together. The *delivery note* itself (see Fig. 6.49) is the final document that you can include in the goods shipment or send to the recipient of the goods electronically.

```
IDES US Inc., 1230 Lincoln Avenue,     Delivery note
10019 New York
Thomas Bush Inc.                       Number/date
1 1 2800 South 25th Ave                80014862 / 05/10/2006
MAYWOOD IL   60153                     Reference number/date
                                       PO789
                                       Order number/date
                                       31129 / 05/10/2006
                                       Customer no.
                                       3000
```

Shipment details

Conditions
Shipping conditions Standard
Terms of delivery CIF Chicago

Weights (gross/net) - Volumes - Selections
Gross weight   61.500 KG Net weight   61.500 KG

Item	Material Description	Quantity	Weight
000010	M-10 Flatscreen MS 1775P PO number_K /Item: PO789	3 PC	61.500 KG

Fig. 6.49 Delivery note in SAP ERP

6.4.2   Shipment

In this section, we take a look at outgoing shipments in SAP ERP based on generated deliveries. The shipment document (see Fig. 6.50) is created by a transportation planning point (which, for instance, corresponds to the Planning Department of your company). A planner has a variety of options with which to select shipments for a transportation document based on various criteria and translate these selections into a shipment. The shipment document acts as the framework for the allocated deliveries. You can set individual transport types (such as road transport) that your company employs. You can also enter in the shipment document which carrier will take charge of the shipments. The shipment document has comprehensive status management functions to monitor the handling of a shipment.

Based on the shipment document, you can have the system calculate the freight charges in the shipment cost document that are accrued for a particular carrier, and thus execute account settlement with the carrier in the classic way (via credit-side settlement) or according to the credit memo procedure. You can also subsequently distribute the values in the shipping cost document to the individual deliveries, to charge your customers the accrued freight charges. Like the sales prices, the freight charges are determined by Pricing depending on a pricing procedure.

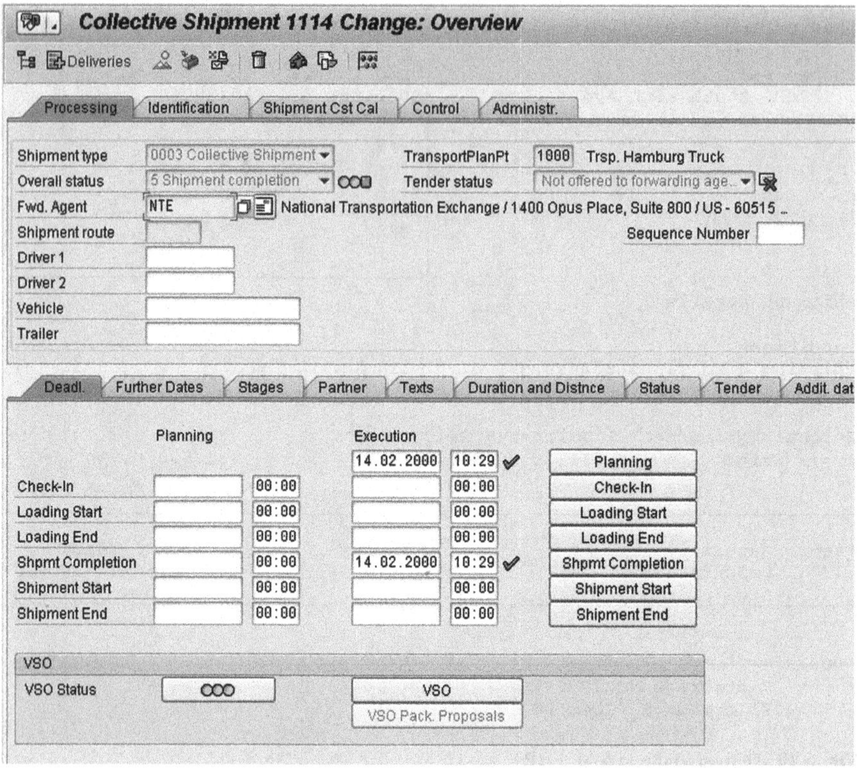

Fig. 6.50 Shipment document in SAP ERP

Information on the Topic of Shipments

You can find detailed information on the topic of Shipment and Transport in SAP ERP and/or SAP SCM (TP/VS or TM) in Volume II, Chapter 2, "Transport Logistics".

6.5   Billing

Billing marks the completion of the logistics process and creates an invoice for your customer. An invoice can be generated from within the ERP system as well as the CRM System (*CRM Billing*). The invoice is based on the logistics documents of sales order or delivery. Once you have created an invoice, the corresponding document flow, Accounting and the statistics are updated. The party to whom you send the invoice is known as the *bill-to party*, and the party that pays the bill is the *payer*. The payer is thus responsible for settlement of the invoice, and is the so-called debitor in the eyes of Financial Accounting. That is the party to whom Financial Accounting will send a reminder in the event of nonpayment (see Fig. 6.51).

The system defines order and delivery-based invoices, as well as credit and debit memos, as possible *billing types*. Examples include a credit memo issued because of defects or a debit memo for additional claims. In Invoicing, in the case of a rebate agreement that you have stipulated with a customer, business partner or employee (such as for field sales), the corresponding amounts are determined and reserved for subsequent rebate payment. Otherwise, Invoicing possesses the same basic functions as the sales order. Thus, you can set how pricing will occur in the event of an invoice generation, that is, whether prices are to be taken from the sales order and taxes recalculated or whether, for example, a completely new pricing calculation is to be run.

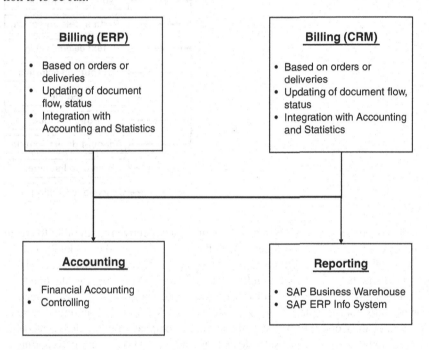

Fig. 6.51 Billing

This is a scenario that you can modify depending on your business needs. Invoicing is also integrated with the organizational structures of sales and distribution, meaning you can trigger invoicing from an area of sales and distribution.

6.5.1 Billing Processing

As we have seen, the invoice can be created on the basis of a sales order (such as a service order), a delivery or a contract (for example, a rental agreement with periodical invoicing) in SAP ERP. The data from the corresponding preceding object (a sales order or delivery) is used in the invoice with the aid of the copy control, which you can set up individually. You can transfer several preceding objects, such as deliveries, into a collective invoice if the criteria allow it. The invoice consists of header and item data, and has basic functions such as the determination of partner, pricing, output and text (see Fig. 6.52).

Fig. 6.52 Billing type and what it influences

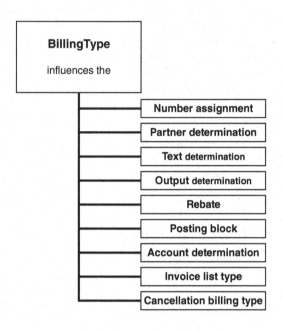

You can create invoices either manually or via a collective run. In order to create an invoice, the preceding documents must have been processed completely and the billing index updated. The billing index is generated, for instance, when a goods issue is booked and the delivery is thus released for invoicing. You also have the option of collecting issued invoices in invoice lists, if you and your customer have agreed to monthly bills, for example. Through seamless integration with Accounting (Financial Accounting and Controlling), saving the invoice also updates the respective accounts for receivables, sales revenue, sales deductions and taxes in Accounting. Reporting is also updated in SAP NetWeaver BW or in SAP ERP itself.

As is the case with all other documents that you create in the system, Billing also offers the possibility of defining printed documents using output technology. The

Billing Options in SAP ERP

Figure 6.53 displays an invoice in SAP ERP and Fig. 6.54 shows an invoice from SAP CRM (CRM Billing).In Fig. 6.53, you see the header detail of the invoice with its accounting and pricing data, as well as taxes and general information.

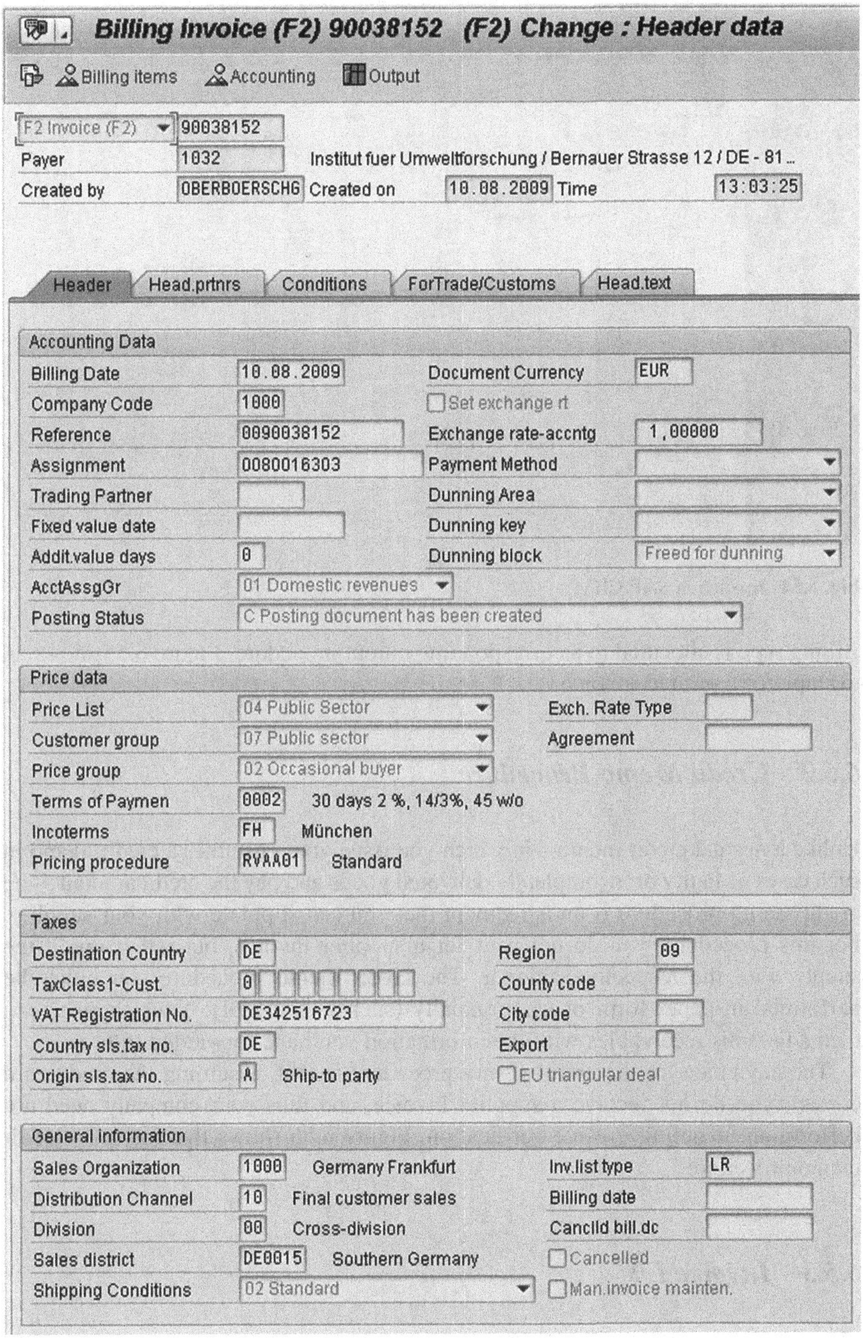

Fig. 6.53 Invoice in SAP ERP

Fig. 6.54 Invoice in SAP CRM

Billing type is allocated to a corresponding output procedure. Figure 6.55 shows an example of a printed invoice.

6.5.2   Credit Memo Procedure

Unlike a normal credit memo—in which you issue your customer a credit memo in such cases as faulty or incompletely delivered goods and pay the credit amount—the credit memo procedure is an agreement that you can stipulate with your supplier. For this procedure, you do not wait for a supplier invoice, but rather credit the supply with the respective amount. The credit memo procedure can either be performed in paper form or electronically (such as via EDI). Your supplier can then adjust his receivables with the information you have forwarded to him.

The advantage of the credit memo procedure is that, assuming the credits are correct, you do not receive a supplier invoice, and thus your company need not perform an incoming invoice verification. Figure 6.56 shows the flow of a credit memo procedure.

6.5.3   Invoice List

If you have made the appropriate agreement with your customer, you can use an *invoice list* (see Fig. 6.57) to summarize several invoices, credit and/or debit memos and send them periodically to your customer.

IDES US Inc., 1230 Lincoln Avenue, 10019 New York

RIWA Regional Warehouse Denver
25 State St.
DENVER CO 80201

Invoice	
Number	90033803
Date	Jun 25, 2003
Reference Number	test 2
Delivery Number	80012661
Date	Jun 25, 2003
Order Number	9112
Date	Jun 25, 2003
Customer Number	6000
Your tax Number	DE342367431

Currency   USD

Conditions:

Terms of payment	Up to 07/09/2003 you receive 3.000 % discount
	Up to 07/25/2003 you receive 2.000 % discount
	Up to 08/09/2003 without deduction
Terms of delivery	FH

Weights - Volume:

Net weight:	4,280 KG
Gross weight:	4,360 KG

Item			Item Detail				
10	Material:	Y-353	Farbe 10 Liter Dose				
	Quantity:	0					
20	Material:	Y-353	Farbe 10 Liter Dose				
	Batch:	0000000206					
	Quantity:	400 KG					
	Prices:	Type		Rate RU	per	CUD	Value
		Gross		51.10 USD	per	1 KG	20,440.00
30	Material:	CH_5103	Lackgrundierung schwarz 2 L Dose				
	Quantity:	0					
40	Material:	CH_5103	Lackgrundierung schwarz 2 L Dose				
	Batch:	0000000207					
	Quantity:	140 KG					
	Prices:	Type		Rate RU	per	CUD	Value
		Gross		35.80 USD	per	1 KG	5,012.00

Items total:						25,452.00

Price Type	Tax	Rate RU	per	CUD	Base	Value
Tax Jur Code Level 1	S1	0.000			25,452.00	0.00
Tax Jur Code Level 1	S1	100.000 %			0.00	0.00
Tax Jur Code Level 2	S1	0.000			25,452.00	0.00
Tax Jur Code Level 2	S1	100.000 %			0.00	0.00
Tax Jur Code Level 3	S1	0.000			25,452.00	0.00
Tax Jur Code Level 3	S1	100.000 %			0.00	0.00

Fig. 6.55 Invoice printout in SAP ERP

A common use of invoice lists is invoicing to buying groups, that is, a central unit (central office) that regulates the invoices of its branches. For instance, the central office receives a monthly invoice list, which includes all of the individual branch invoices during the cited period, and can trigger settlement of the invoices in

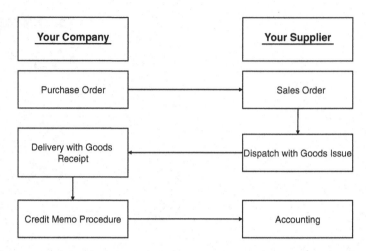

Fig. 6.56 Credit memo procedure

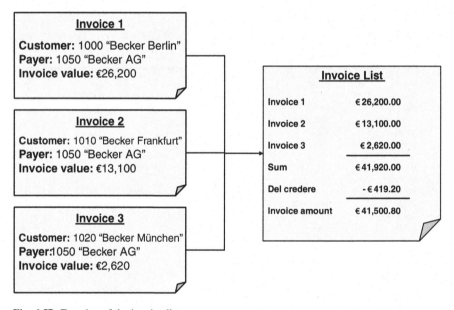

Fig. 6.57 Function of the invoice list

this way. The customer can stipulate a *del credere* (guarantee of ability to pay) with you. Generally, a percentage of the del credere is deducted from the invoice value, since the central regulator takes on the guarantee for payment in the name of the actual debtors (in this case, the branch) and receives payment for it. The advantage to your company is that the payment of the debt is guaranteed. Furthermore, the invoice list (see Fig. 6.58) is a clear instrument for the central office that provides an overview of branch purchasing.

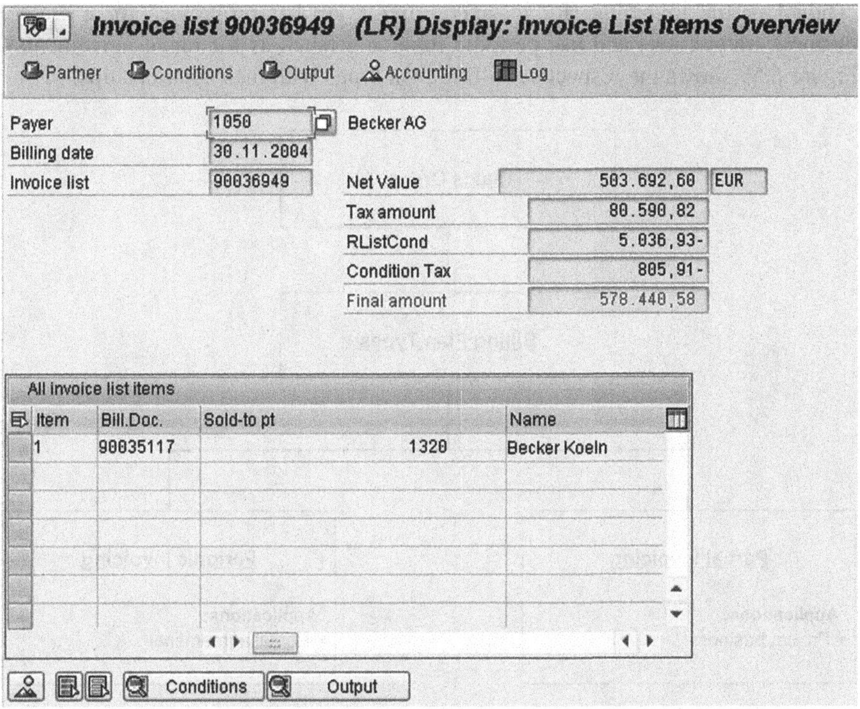

Fig. 6.58 Invoice list in SAP ERP

Invoice List in SAP ERP
Figure 6.58 displays an invoice list to Payer 1050 (Becker AG) with Invoice 90035117 for Customer 1320 (Becker Koeln) with a net value of €503,692.60. In this case, the del credere is determined from the net value and tax amount.

As depicted in Fig. 6.58, an invoice list can contain any amount of individual invoices. The central payer receives an invoice list on certain dates that you define in a customer calendar, which contains the corresponding individual invoices. Thus, the central office has a clear instrument for processing goods and services invoices of its affiliates and receives an overview of outstanding payments of the respective branches.

6.5.4 Billing Plan

You can save a *billing plan* in the sales order. This is a good idea if you bill your customers for periodic services. In the ERP system, you have the opportunity to conduct partial invoicing (for instance, after a job has progressed for a project business transaction) and use periodic invoicing (such as for rental agreements). Figure 6.59 shows the respective billing plan types with their relationships.

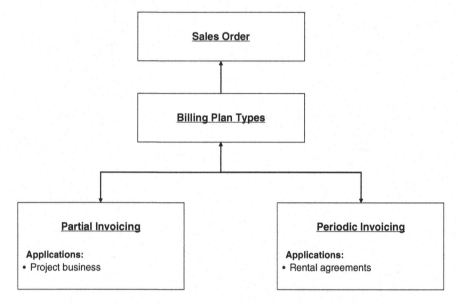

Fig. 6.59 Billing plan types

Partial invoicing is used when the total value of individual milestones, such as for project business, is billed. This type of *periodic invoicing* is performed when the amount to be charged will be invoiced at periodic times, such as for rent (see Fig. 6.60).

6.5.5   Rebate Processing

With *rebate processing*, you can grant your customers a direct or subsequent rebate (a special kind of discount) depending on attained turnover within a defined period. The basic elements of rebate processing are depicted in Fig. 6.61.

You can define the type of rebate to be granted in the *rebate agreement*. There, you can save data on the recipient of the rebate and the rebate criteria to be used. When an invoice is posted (individual invoice, credit or debit memo), the system uses the rebate agreement to determine the estimated rebate amount and posts reserves for that amount.

Based on the reserves, your Financial Accounting Department has an updated overview of the expected rebate payments. At the end of the stipulated period, you can release the rebate agreement for settlement.

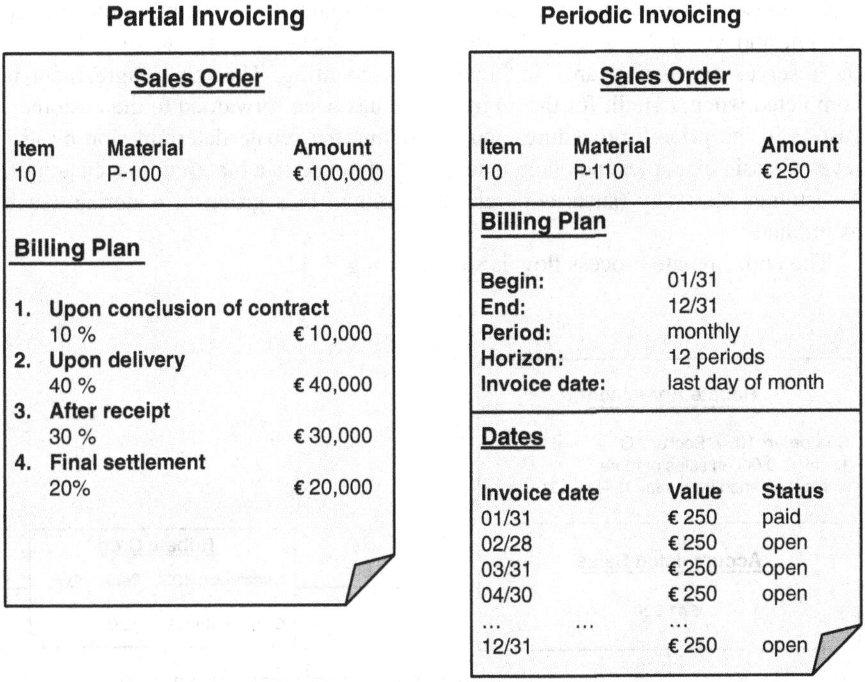

Partial Invoicing
Sales Order
Item Material Amount
10 P-100 € 100,000
Billing Plan
1. Upon conclusion of contract
10 % € 10,000
2. Upon delivery
40 % € 40,000
3. After receipt
30 % € 30,000
4. Final settlement
20% € 20,000
Periodic Invoicing
Sales Order
Item Material Amount
10 P-110 € 250
Billing Plan
Begin: 01/31
End: 12/31
Period: monthly
Horizon: 12 periods
Invoice date: last day of month
Dates
Invoice date Value Status
01/31 € 250 paid
02/28 € 250 open
03/31 € 250 open
04/30 € 250 open
... ...
12/31 € 250 open

Fig. 6.60 Partial invoicing vs. periodic invoicing

Define rebate agreement
Accumulate revenue and accruals
Release rebate agreement for settlement
Issue rebate agreement for settlement and cancel accruals

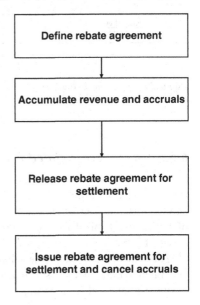

Fig. 6.61 Overview of rebate processing

For settlement of the rebate agreement, the system determines the volume rebate amount and generates a credit (rebate credit memo). When the credit is posted, the reserves are also cleared in Financial Accounting. The rebate agreement is completed when a credit for the entire rebate has been forwarded to the customer. Just as in the pricing procedure, you can define the rebate determination on any level. Classic examples include rebates with reference to a material, to a customer, a customer hierarchy (such as purchasing groups) or a group of materials (such as product lines).

The entire rebate process flow is shown in Fig. 6.62.

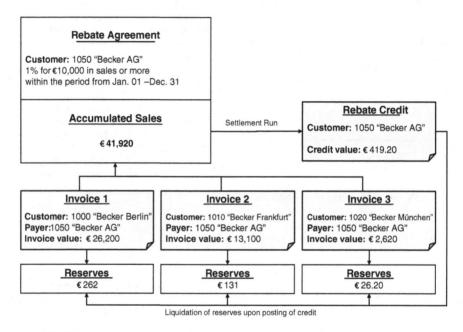

Fig. 6.62 Rebate processing flow

6.5.6   Resource-Related Billing

Resource-related billing is used when the prices for a provided service cannot be entered into the sales order as a fixed value, but are determined by the expenditure of resources. Typical examples include project business, production to order or service business.

You can invoice such orders based on resources, meaning that the invoice is generated based on the "subsequently calculated" sales order. The function of resource-related billing is depicted in Fig. 6.63.

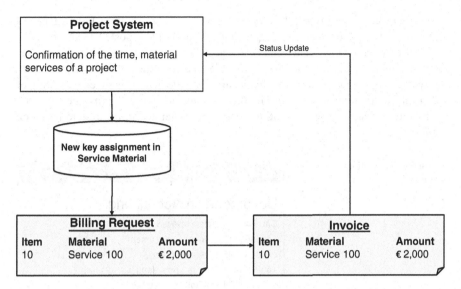

Fig. 6.63 The resource-related billing process

6.5.7 Effects of Invoice Generation

The invoice that you create in Sales and Distribution forms the basis for the corresponding document in Accounting and the updating of other documents (see Fig. 6.64).

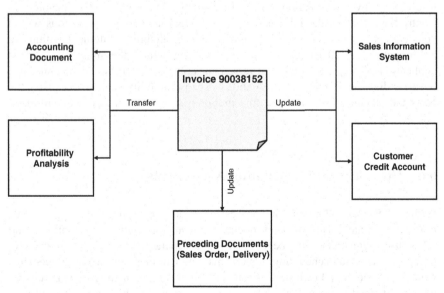

Fig. 6.64 Effects of invoice generation

Generation of the invoice has various effects on Accounting: For instance, a debit is posted to the receivables account and a credit is posted to the revenue account of the customer. The data of the accounting document is based on information from the pricing of the invoice itself. You can set the transfer of invoice data to Accounting so that it automatically occurs when the invoice is saved. Updating of other information, as depicted in Fig. 6.64, also occurs when the invoice is saved.

Figure 6.65 shows the accounting documents generated in the system based on the invoice.

Fig. 6.65 Accounting documents

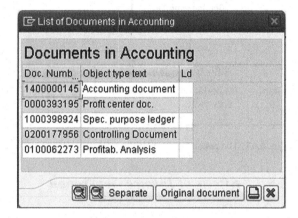

The accounting document in Financial Accounting serves as proof of value-based changes within your company. These value-based changes can occur as a result of an invoice based on a sale, for example. The profit-center document in Controlling contains the information of individual, autonomous divisions within your company and serves as the basis for the profitability statement within the profit center. Special ledgers aid you in reporting values stemming from various applications. The cost accounting document collects the costs and revenue of a business division to calculate profitability. The profitability analysis compares the costs and revenues upon which a contribution margin account can be determined.

6.6   Contracts and Scheduling Agreements

Within SAP sales and distribution processing, contracts and scheduling agreements represent special forms of sales documents. Quantity and value contracts and scheduling agreements can be defined via a separate type of sales document. Using contracts and scheduling agreements, you can map contractual agreements with your customers. For instance, you can save a certain quantity to be purchased within a defined period, such as 1 year, in a contract. In this case, it would be a quantity contract. During the order entry, the system checks whether a contract exists for that particular configuration. If so, the relevant information is taken from it and the contract is updated from the order, which in the case of a quantity contract leads to a reduction in the contractual quantity. Using contracts, you can achieve better planning for your company, because it is contractually fixed. We generally differentiate between contracts and delivery scheduling agreements (see Fig. 6.66).

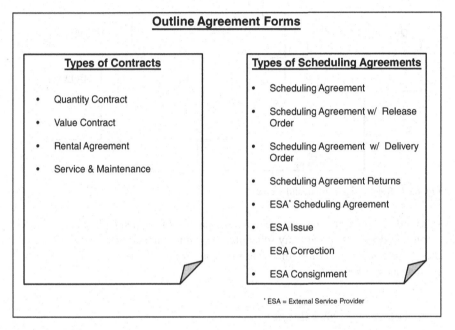

Fig. 6.66 Types of contracts

In the following sections, we will take a closer look at quantity and value contracts as well as scheduling agreements.

6.6.1   Quantity Contracts

Quantity contracts are used to determine for your customer what quantities of one or more products are to be purchased within a certain period. If you receive an order from a customer (release order), the system checks the corresponding contract and updates the release quantity in the quantity contract. Thus, you always have an overview of the contractually stipulated quantity, the released portion, and the remaining quantity. Furthermore, you can set the system such that, when a certain quantity has been reached, a subsequent action is triggered that informs you that the contract will be exhausted shortly and that you should initiate an extension or renegotiation of the contract. Figure 6.67 shows a quantity contract and the corresponding sales orders (releases), as well as the respective deliveries. The orders and order quantities lead to an updating, that is, a reduction, of the contractual quantity.

In Fig. 6.68, you can see the system document of the quantity contract. Like a

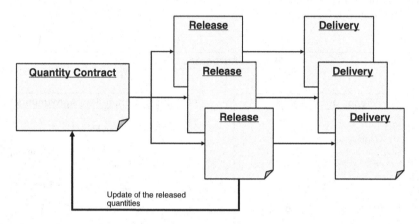

Fig. 6.67 Release against a quantity contract

sales order, a quantity contract can contain one or more items. Unlike a sales order, a contract does not contain any schedule lines, since they are included in the sales orders (releases). Generally, the same functions are available in a quantity contract as in a sales order.

Change Quantity Contract 40000089: Overview

Quantity Contract	40000089	Net value	1.000.000,00	USD
Sold-To Party	1470	Aircraft Products / 185 Farnham Road / Slough		
Ship-To Party	1470	Aircraft Products / 185 Farnham Road / Slough		
PO Number	ITARContract	PO date		

Sales | Item overview | Item detail | Ordering party | Procurement | Shipping | Reason for rejection

Description	
Valid from	01.10.2003	Valid to	01.10.2005
Billing block		Pricing date	01.10.2003
Order reason	
Sales area	3000 / 10 / 00	USA Philadelphia, Final customer sales, Cross-division
Master contract	
Shipping Cond.	10 Immediately
Business Area	

All items

Item	Material	Target quantity	U...	Description	Customer Material Numb
10	6TS-020	1.000	EA	Aircraft Special Control Part	

Fig. 6.68 Quantity contract in SAP ERP

The Quantity Contract in SAP ERP
Figure 6.68 shows a quantity contract for Customer 1470 (Aircraft Products). This quantity contract stipulates that your customer plans to purchase 1,000 units of Material GTS-020 ("Aircraft special control part") in the period from Oct. 1, 2003 to Oct. 1, 2005.

The quantity contract contains the quantity to be purchased within a stipulated period and agreed prices.

6.6.2   Value Contracts

Value contracts (see Fig. 6.69) are used to stipulate with your customer a contractual agreement for a certain value within a defined period. Your customer is then obligated to purchase products and/or services from you adding up to the value stipulated in the contract.

Fig. 6.69 Structure of a value contract

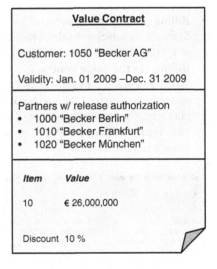

The releases (orders) against a value contract can be related to individual products or services or a group of products, such as a product line (see Fig. 6.70). Furthermore, you have the opportunity to add *partners with release authorization* to the contract. This can be especially useful if, for example, the contract is with an affiliated group (see Sect. 6.2.4) and the group's subsidiaries are also authorized to release.

If releases against a value contract are to take place in more than one currency, the contract is updated by converting the currencies to the contract currency. Invoicing of the value contract can either be done via a release (order) or directly:

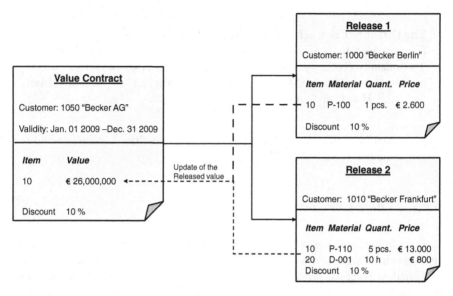

Fig. 6.70 Releases against a value contract with several authorized partners

- **Billing per release order**
  Since the releases are sales orders, you can either bill them with relation to an order or a delivery (see Sect. 6.5.1).
- **Billing via the value contract**
  When you invoice a value contract, it is usually done with reference to an order, meaning you invoice the value contract itself. If you have stipulated a payment plan with your customer, you will use a billing plan (see Sect. 6.5.4), where you can save the billing modalities, such as the existence of several billing periods or value percentages.

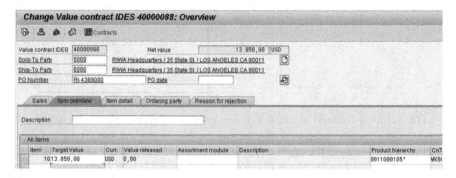

Fig. 6.71 Value contract in SAP ERP

In addition, you can combine individual contracts into group contracts. The data from the group contract then applies to all allocated contracts (subcontracts). If you change the data of the group contract, the subcontract data is also updated.

Value Contracts in SAP ERP
Figure 6.71 shows a value contract for Customer 6000 (RIWA). It stipulates that the customer will make purchases having a contractual value of €13,850. Releases (orders) are posted against that value in the system.

6.6.3   Scheduling Agreements

A *scheduling agreement* is primarily used in the component supply industry. In it, you can define set delivery quantities and dates (see Fig. 6.72). That is why, unlike other types of contracts, a scheduling agreement already has schedule lines to be used for production or external procurement and to create a delivery. When schedule lines in the delivery schedule become due, you can create deliveries either directly or via the delivery due list. For more information on creating deliveries, see Sect. 6.4.1.

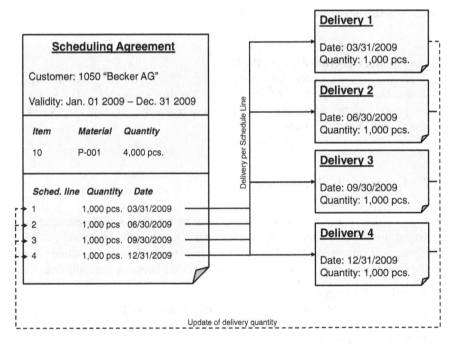

Fig. 6.72   Scheduling agreement and the respective deliveries

When schedule line entries are made for an item of the delivery schedule (see Fig. 6.72), the system adds up quantities already entered and compares them with the quantities already delivered. You thus have a continuous overview of open delivery quantities. If your customer wishes to receive a periodic invoice, for instance on a monthly basis, you can combine all deliveries due in a single, collective invoice.

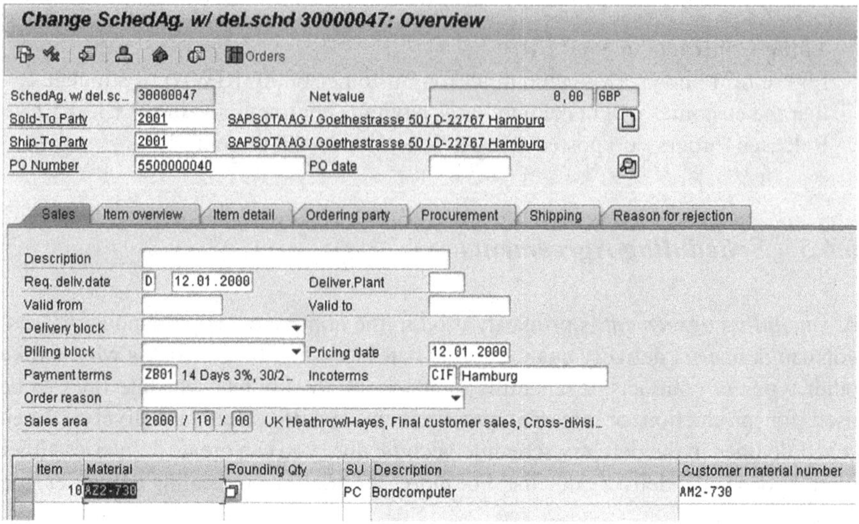

Fig. 6.73 Scheduling agreement in SAP ERP

Scheduling Agreements in SAP ERP
Figure 6.73 shows a scheduling agreement for Customer 2001 (SAPSOTA).
It stipulates the delivery of Material AZ2-730 ("Bordcomputer"). The validity period can be indicated by supplying the respective dates in the "Valid
from" and "Valid to" fields.

6.7   Complaints Processing

You can utilize *complaints processing* if your customer has received defective,
damaged, insufficient, incorrectly ordered or incorrectly delivered goods, if price
agreements are incorrect or if returns arrive before the goods return deadline passes.
In the case of complaints processing, we differentiate between the following:

- **Credit memo request**
  For this, you grant your customer a refund without verification. This is done
  when, for instance, you have charged the customer the wrong price.
- **Debit memo request**
  In this case, the opposite of the credit memo request occurs. The debit memo
  request is used if you have not charged your customer enough.
- **Returns with free-of-charge subsequent delivery or credit memo**
  You create a return when your customer sends goods back to your company.
  With the return, you document the process of receiving the defective goods.
  Based on the returns document, you can issue the customer a free-of-charge
  subsequent delivery or refund.

- **Invoice correction request**
  To place an invoice correction request, enter the quantity or price that should have been invoiced. Based on this amount, the system calculates the difference. If it is negative, a debit memo is generated; otherwise, a credit memo is generated.

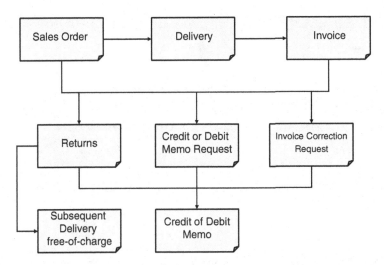

Fig. 6.74 Complaints processing procedures

Figure 6.74 demonstrates the procedures involved in processing complaints.

You can trigger complaints processing from the ERP or CRM system. Figure 6.75 shows complaints processing in SAP ERP, and Fig. 6.76 depicts the procedure in SAP CRM.

Complaints Processing in SAP ERP
Figure 6.75 presents an overview of complaints processing. In this case, you have to process a complaint from Customer 1032 (Becker AG) regarding Delivery 90038152. For this, you create a complaint document in which the reason for the complaint is recorded.

6.7.1   Credit and Debit Memo Requests

A *credit or debit memo request* can be created if there are grounds for price differences or if goods are to be credited without being returned.

- A credit memo request is used if you wish to credit your customer with a certain value.
- A debit memo request is used when you wish to bill a customer for an additional service that has not been invoiced in advance.

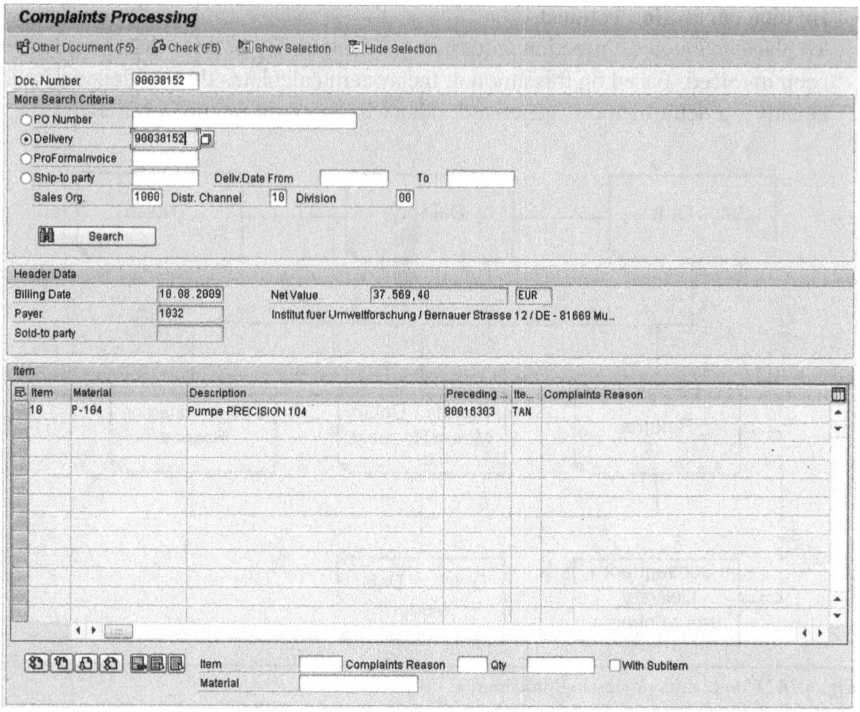

Fig. 6.75 Complaints processing in SAP ERP

Fig. 6.76 Complaints processing in SAP CRM

As Fig. 6.77 shows, you can either create credit or debit memo requests with or without reference to a sales order or invoice. The data from the preceding document is used to create such a request with reference to the respective document, and can be revised directly in the requests. This procedure ensures that the amounts for which a credit or debit memo is required are correctly adopted.

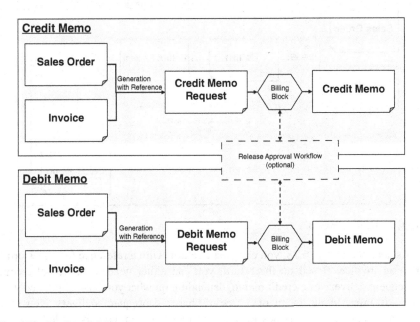

Fig. 6.77 Credit and debit memo request process

By default, the system generates a billing block, in order to guarantee that credit and debit memo requests are checked by an authorized employee of your company before the actual credit or debit memo is issued. Optionally, this check can be forwarded via a workflow to an authorized employee for approval, depending on the value amount. As soon as that employee has granted approval, the billing block is removed and the actual credit or debit memo is generated. If the request is not approved, you can record a reason for the rejection in the request. The actual generation of the credit or debit memo is done in the same way as the generation of invoices, described in Sect. 6.5.

6.7.2 Returns

Use the *Returns* function if your customer sends goods back to your company. The returns document and the return delivery are the basis of posting the goods, for example, to the inspection stock of the warehouse. This is the area where an inspection of the returned goods is performed, to determine if they may be sold again or, in the worst case, need to be scrapped.

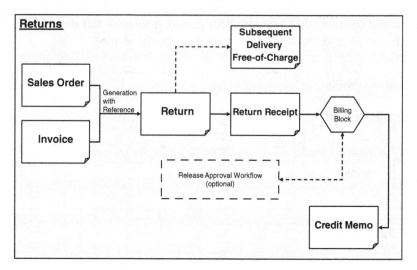

Fig. 6.78 Returns process

As depicted in Fig. 6.78, you can create returns with a reference to a sales order or to an invoice. Based on the return, you can either generate a free-of-charge subsequent delivery or a credit memo, depending on what you have stipulated with your customer. In the event of a free-of-charge subsequent delivery, you send your customer the same goods again, such that the complaint is settled with an "exchange". However, if your customer wants a refund, you also create a credit memo. Settlement of this credit memo is done as described in the previous section, "Credit and debit memo requests".

Returns represent return deliveries of goods from your customer that you check in goods receipt and, if the goods are in order, post to stock. For more information on the goods receipt process, see Chap. 4, Procurement Logistics, and Volume II, Chapter 3, "Warehouse Logistics and Inventory Management".

6.7.2.1   Subsequent Delivery Free-of-Charge

One way to settle a returns process is through the "subsequent delivery free-of-charge" of the goods (see Fig. 6.80). Based on the return, you create a subsequent delivery free-of-charge (a special sales document type). In this document, you enter the reason for the subsequent delivery (order reason) and generally a billing block. Subsequent deliveries free-of-charge with a billing block are checked by an employee and, if a subsequent delivery is to take place, the billing block is removed and the document is due for shipment. For more on delivery processing, see Sect. 6.4.1.

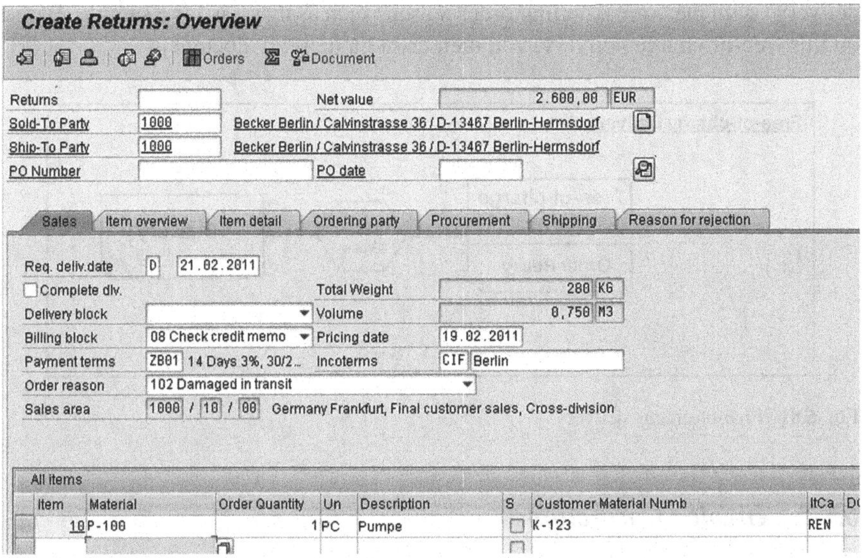

Create Returns: Overview
Orders
Document
Returns
Net value
2.600,00
EUR
Sold-To Party
1000
Becker Berlin / Calvinstrasse 36 / D-13467 Berlin-Hermsdorf
Ship-To Party
1000
Becker Berlin / Calvinstrasse 36 / D-13467 Berlin-Hermsdorf
PO Number
PO date
Sales
Item overview
Item detail
Ordering party
Procurement
Shipping
Reason for rejection
Req. deliv.date
D
21.02.2011
Complete dlv.
Total Weight
280
KG
Delivery block
Volume
0,750
M3
Billing block
08 Check credit memo
Pricing date
19.02.2011
Payment terms
ZB01
14 Days 3%, 30/2...
Incoterms
CIF
Berlin
Order reason
102 Damaged in transit
Sales area
1000 / 10 / 00
Germany Frankfurt, Final customer sales, Cross-division
All items
Item
Material
Order Quantity
Un
Description
S
Customer Material Numb
ItCa
10
P-100
1
PC
Pumpe
K-123
REN

Fig. 6.79 Overview of returns in SAP ERP

Returns in SAP ERP
Figure 6.79 shows a returns document for Customer 1000 (Becker Berlin). The return is recorded with the reason "Damaged in transit" and the billing block "Check credit memo".

Subsequent Delivery Free-of-Charge
Return
Subsequent delivery free-of-charge
Order Reason
Delivery Block
Delivery

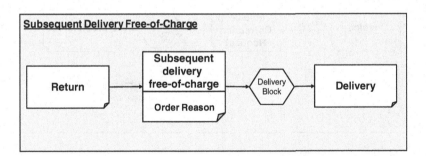

Fig. 6.80 Subsequent Delivery Free-of-Charge based on a return

6.7.2.2 Free-of-Charge Delivery (Special Case)

You can use a free-of-charge delivery (see Fig. 6.81) if you wish to provide customers with a sample. To do this, create a document of the type "free-of-charge

delivery" with the corresponding order reason, such as that it is for a sample. Based on this free-of-charge delivery, you then create a delivery document.

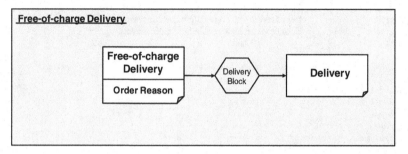

Fig. 6.81 Free-of-charge delivery

6.7.3   Invoice Corrections

In the case of an *invoice correction*, you adjust quantities or values for which the customer has already been billed. As depicted in Fig. 6.82, the invoice correction request is always created with reference to an invoice. As a result, your customer receives a document listing the originally billed quantities and values as well as the corrected quantities and values used for a credit memo.

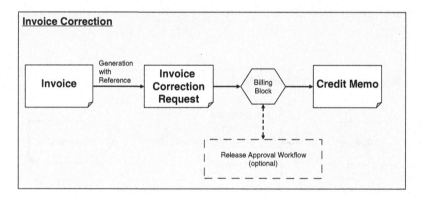

Fig. 6.82 Invoice correction process

6.7.4   Blocking and Releasing Complaints

As we have seen, complaint documents have corresponding *blocking mechanisms* designed to prevent you from processing a complaint further without having it reviewed first. The following blocking mechanisms are especially important:

- **Delivery block**
  A delivery block is set for free-of-charge deliveries and subsequent deliveries to prevent goods from leaving the plant without being calculated. In order to lift the delivery block, you have to possess the respective authorization.
- **Billing block**
  A billing block is set for documents that perform a value correction (such as a credit memo), since you generally want to check the documents based on a complaint that lead to a credit memo—in other words, a material outflow from your company. To lift the billing block, you need the respective authorization.

As soon as the respective blocks have been lifted, the documents can be processed further and follow-on documents generated.

6.7.5 Rejecting Complaints

Unjustified complaints that you are reviewing can be rejected by entering a reason for rejection (see Fig. 6.83).

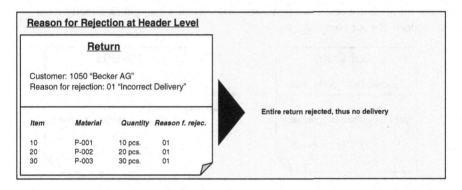

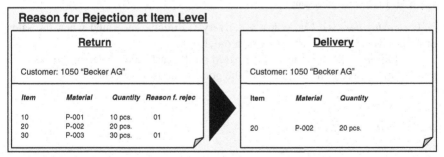

Fig. 6.83 Reason for rejection using the example of a return

You always have an overview of the open and rejected complaints in the system. You can either reject the entire complaint document or individual items of the business transaction. If you set the reason for rejection on the item level, only those items that are not rejected are taken on in the follow-on document. Setting a reason for rejection for free-of-charge deliveries or subsequent deliveries results in the document items not being able to be delivered. In other complaint cases, the result is that no credit or debit memo or invoice correction can be generated.

6.8   Special Business Transactions in Sales and Distribution

In this section, we explain special business transactions in Sales and Distribution, such as rush orders and cash sales. Such business transactions build on the structures described in the previous sections, such as the sales order.

6.8.1   Rush Order and Cash Sales

A *rush order* (see Fig. 6.84) differs from a standard sales order in that as soon as the sales document is saved, a delivery is directly generated. You can use this type of order for such instances as when your customer picks up goods directly from the sales desk of your company.

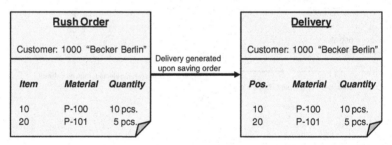

Fig. 6.84 Rush order processing

For *cash sales* (see Fig. 6.85), a delivery is directly generated when the sales order is saved, as is the case for a rush order.

In addition, you create an invoice based on the cash sales (sales order) because you not only directly trigger issue of the goods but also payment.

6.8.2   Individual Orders

In the case of an *individual order*, a purchase requisition is directly generated from the sales order schedule line (see Fig. 6.86). This type of processing is used for materials that you never have in stock and are always procured externally. This could be the case if, for instance, your customer has ordered a specific material that your company does not produce and always purchases on the basis of sales orders.

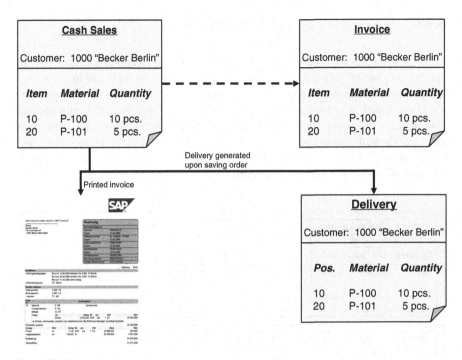

Fig. 6.85 Cash sales processing

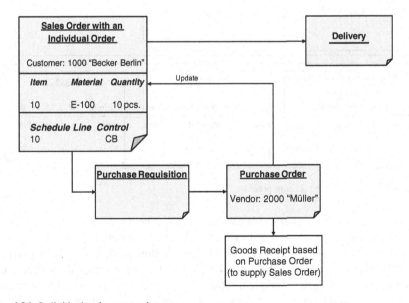

Fig. 6.86 Individual order processing

Based on the purchase requisition generated from the sales order, Purchasing generates an order in your company and a specific vendor. The goods are directly delivered to your company, where they are then available for the supply of the sales order.

6.8.3   Returnable Packaging Processing

Returnable packaging processing refers to goods at your customer location that are still the property of your company. Your company is authorized to bill the customer for the provided (loaned) goods if they are not returned to you. Returnable packaging is processed as special stock in your inventory management. This ensures that you process inventory separately and always know what returnable packaging (material) is with which customer, since you maintain special stock separately for each customer. A common usage of this type material in logistics is the euro-pallet. Figure 6.87 shows the general procedure for returnable packaging.

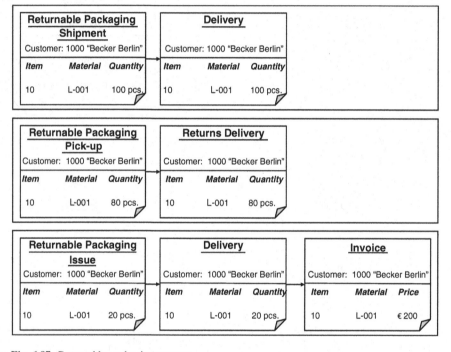

Fig. 6.87 Returnable packaging process

You first generate a *returnable packaging shipment* (a type of sales order). Based on that, you create a delivery with which you post the materials from your stock to customer special stock and deliver them to the customer. If your customer reports that the goods or parts of them can be picked up, you then create a *returnable packaging pick-up* and subsequent return delivery. Based on these, the returnable packaging is picked up from the customer and placed into your warehouse. When the goods issue is posted into the warehouse, the goods are transferred from special stock to unrestricted stock. In order to invoice goods that remain on the premises of the customer, you need to create a *returnable packaging issue* with a subsequent delivery and invoice. The delivery, with its goods issue posting, clears the goods from special stock. Physical transfer is not triggered, since the customer already has the goods.

> **Third-Party and Consignment Processing**
> The topics of third-party and consignment processing are examined in Chap. 4, "Procurement Logistics".

6.9 Summary

In this chapter, we have presented the business significance and tasks of distribution logistics. The goal of the chapter has been to make you familiar with the systems, components and applications of SAP Business Suite and its tasks and functions in the sales process, from the sales order to invoicing.

The purpose of sales and distribution is to sell the produced or procured goods via an efficient distribution channel. Sales and distribution is mapped in the ERP system in the component SD, and sales and distribution processes are deeply integrated in the procurement, production, warehouse and invoicing processes of SAP ERP. This chapter has referred to the entire process flow, from the customer inquiry to the quotation, from the sales order to delivery and invoicing. In addition, we have highlighted the key basic functions, such as pricing and special business transactions like rush orders, to provide you with a general overview of the processes and functions in SAP systems for distribution logistics.

This volume has presented the functions and tools of SAP Business Suite for the logistical execution of supply chain fulfillment. Following a general explanation of data organization, we turned to the organizational structures pertaining to procurement logistics, citing applications for such tasks as requirements determination, purchasing, order processing, and goods movement. SAP systems and components for production logistics were then introduced, including those for demand and production planning and detailed scheduling. Finally, the role of SAP applications in distribution logistics was detailed within the context of sales, invoicing, contracts and scheduling agreements, complaints and special distribution needs.

Volume II of this series, *Inventory Management, Warehousing, Transportation and Compliance*, expands on the scope of SAP's role in supply chain execution, covering the realms of transport logistics and its integration in procurement and distribution logistics. It also explains SAP processes involved in warehouse and inventory management and foreign trade formalities. Finally, it examines ways in which SAP tools can aid you in tracking objects, managing exceptions, tracing processes and generating intuitive reports.

Glossary

ABAP The programming language of SAP with which most logistics applications are programmed.

Activity area A special feature of SAP EWM representing a logical grouping of storage bin locations with regard to planned warehouse activities.

ALE Application Link Enabling. Technology that establishes and operates shared applications.

APO Advanced Planning & Optimization. SAP APO software contains functions for processing and integrating sales, distribution and production planning, as well as production control and external procurement. SAP APO also offers functions for collaboration with external vendors and their integration in procurement processes (see also VMI).

APO PP/DS Production Planning/Detailed Scheduling. An SAP APO module enabling the planning of production within a factory while simultaneously considering product and capacity constraints, with the goal of increasing throughput and reducing product stock. The result of the planning is a feasible production plan.

ATP Available-to-Promise. Refers to warranted stock, the quantity of a certain material that can be made available at a required date or subsequent time and can thus be used for such purposes as sales orders. The system takes into account the current stock situation and planned inward and outward movements, especially based on purchase orders, production orders and recorded sales orders.

BI Business Intelligence. Business analysis processes and technical instruments in a company used to evaluate companywide data and to provide that data to users.

Business object The copy of a document necessary for a business process in a software system (for example, a sales order in order processing).

CIF Core Interface. Interface for data transfer between an ERP system (SAP R/3 or SAP ERP) and a connected SCM system such as SAP Advanced Planning & Optimization (SAP APO) or SAP Supply Network Collaboration (SAP SNC).

Consolidation The consolidation of goods from several forwarders in a common loading unit (such as a container). Consolidation is executed by a logistics service provider.

J. Kappauf et al., *Logistic Core Operations with SAP*,
DOI 10.1007/978-3-642-18204-4, © Springer-Verlag Berlin Heidelberg 2011

CRM Customer Relationship Management. Supports all customer-related processes within the customer relationship cycle, from market segmentation, lead generation and opportunities to post-sales and customer service. Includes business scenarios such as Field Sales and Service, Customer Interaction Center and Internet Sales and Service.

CTP Capable-to-Promise. Function of the global ATP availability check in which, in contrast to ATP, not only available stock is considered, but also additional sources of requirement coverage, such as production capacities or external suppliers.

Dashboard Display of key logistics data using diagrams and other graphic elements (such as "speedometer"-like displays as on a car dashboard).

EDI Electronic Data Interchange. Cross-company electronic data exchange between business partners (for example, exchanging trade documents).

EHS Environment, Health and Safety. SAP application component for all tasks pertaining to labor, health and environmental safety in a company.

EM Event Management. SAP application component in SAP Supply Chain Management for the monitoring of logistics and other processes.

Embargo list List of people or companies to which it is prohibited to supply certain goods or services.

EPC Electronic Product Code. Code used in RFID chips for product characteristics and identification numbers that is internationally uniform.

ERP Enterprise Resource Planning. The core system with SAP business applications in the areas of logistics, human resources and finances.

Event Handler Generic object in Event Management used to track the status of processes or physical objects (such as shipments tracking).

EWM Extended Warehouse Management. New warehouse management module based on SAP SCM.

FCL Full Container Load. Transport of a full container from a forwarder to a recipient.

Freight invoice Invoice from a logistics service provider to a shipper or recipient of goods that the logistics service provider was commissioned to transport.

gATP global Available-to-Promise. Global availability check with SAP APO on several levels. In contrast to ATP, gATP can check the stock situation of several plants.

GTS Global Trade Services. Component of the SAP system landscape used to process export transactions and consider trade regulations.

HAWB House Air Waybill. Shipment-based house waybill for air freight shipments that is issued by the logistics service provider for the forwarder.

House B/L House bill of lading. Shipment-based house waybill for sea shipments that is issued by the logistics service provider for the forwarder.

HU Handling Unit. Physical unit of packing materials (loading equipment/packaging materials) and the materials kept on or in it. A handling unit has a specific, scannable identification number via which handling unit data can be retrieved.

IMG Implementation Guide. Tool for customer-specific tailoring of SAP systems. The guide has a hierarchical structure based on the application component hierarchy. Central components include the IMG activities, which serve to ensure branching into Customizing and thus the execution of relevant system settings. The following IMG variants exist: SAP Reference IMGs, Project IMGs, and Project view IMGs.

Independent requirement A requirement that is created through a direct influence (for example, material requisition for production or a sales order for the material).

KPI Key Performance Indicator. Key performance figure determined from business transactiondata.

LCL Less Than Container Load. General cargo shipment in which a forwarder's goods are packed into a container with goods of other forwarders (consolidation).

Lead Pick-ups or movementsof goods that are transferred from a local to a long-distance transportation network.

LES Logistics Execution System. SAP application component in SAP ERP with which shipping and transport procedures can be processed.

Letter of credit Documented promise of credit issued by an importer's bank to effect payment to an exporter when the exporter is in possession of the proper documents for an export transaction.

LO General logistics module of SAP ERP.

Master B/L Master Bill of Lading. Consolidation-based waybill that a logistics service provider receives for a consolidated shipment (for example, for several shipments in a single container).

Materials planning Distribution of orders and allocation and provision of resources for the planning of processing.

MAWB Master Air Waybill. Consolidation-based waybill for air freight that an airline issues for the entire cargo of a logistics service provider.

MM Materials Management. Materials management module of SAP ERP.

OER Object Event Repository. Tracking system for RFID-supported logistics processes based on SAP Event Management.

On-carriage Deliveries or shipments of goods that are transshipped from long-distance to local transport.

Operative Planning Planning based on short-term values and goals, such as a transportation plan for the coming day.

Optimization Using targeted methods to find a favorable solution for a complex mathematical or logistics problem. The method generally involves an optimization algorithm (computer program). One example is complex transportation planning.

qRFC queued Remote Function Call. Extension of the transactional remote function call with the option of setting the call sequence.

Requirement The required quantity of a material at a certain time for a specific plant.

RFC Remote Function Call. Calling a function module in a different system (destination) than the one in which the invoked program is running. Connections are possible between various AS ABAP systems or between an AS ABAP and an external system. In external systems, instead of function modules, specially programmed functions are invoked whose interface simulates a function module. There are synchronous, asynchronous and transactional RFCs. Activation of the invoked system is done via the RFC interface.

SAP NetWeaver An open integration and application platform for all SAP solutions and certain solutions from SAP partners. SAP NetWeaver is a Web-based platform that serves as the basis for Enterprise Services Architecture (ESA) and enables the cross-company and technology-independent integration and coordination of employees, information and business processes. Thanks to open standards, information and applications from practically any source can be integrated and can be based on virtually any technology. SAP NetWeaver includes functions for business intelligence, company portals, exchange infrastructure, master data management, mobile infrastructure and a Web application server.

SCM Supply Chain Management contains functions for planning, execution, coordination and collaboration in the supply chain. Among other elements, it is composed of the components and applications APO (Advanced Planning & Optimization), SNC (Supply Network Collaboration) and EM (Event Management). SCM is part of SAP Business Suite.

SCOR Supply Chain Operations Reference Model. Important key data that enables an analytical evaluation of the entire logistics chain.

SD Sales and Distribution. Sales module in SAP ERP.

SNC Supply Network Collaboration. SNC enables the connection of external suppliers to SAP SCM.

SSCC Serial Shipper Container Code. Number of a shipping unit for the identification and labeling of shipping units. A shipping unit under this code is the smallest physical unit of goods and commodities that is not attached to another unit and is or can be treated individually in the transport chain.

Standard software Group of programs that can be used to edit and solve a series of similar or uniform tasks. SAP Business Suite is standard business software.

Strategic planning Planning based on long-term values and goals, such as location planning for production plants.

Tactical planning Planning based on medium-term values and goals, such as production planning for Christmas business.

TM SAP Transportation Management, the transport solution within SAP SCM.

TP/VS Transport Planning/Vehicle Scheduling. Transport optimization in SAP SCM.

Transport request Request of a shipper to a logistics service provider to execute the transportation of goods.

VMI Vendor Managed Inventory. Supplier-controlled inventory for which the supplier has system access to a company's warehouse stock and demand data. The VMI thus enables close cooperation with the supplier and serves to improve the external procurement process.

Warehouse order Generally represents an executable work package to be performed by a warehouse employee within a certain period. Warehouse orders usually consist of the warehouse tasks allocated to them.

Warehouse task A document in SAP EWM containing all necessary information for the movement of a certain material quantity or handling unit in the warehouse.

WM Warehouse Management. A system to define and manage complex warehouse structures within one or more plants. The warehouse management system MM-WM supports warehouse management as well as the execution of all warehouse movement, such as goods put away, removal and transfer.

Bibliography

PL Study (2009) The state of logistic outsourcing 2009 third-party logistics. http://www.uk.
  capgemini.com/services/ceo-agenda/
  the_state_of_logistics_outsourcing_2009_thirdparty_logistics/
Bradler J (2009) SAP supplier relationship management. SAP PRESS, Bonn
Council of Supply Chain Management Professionals (2009) http://cscmp.org/aboutcscmp/
  definitions.asp. Accessed 9 Dec 2009
Gau O (2010) Praxishandbuch Transport und Versand mit SAP LES, 2nd edn. SAP PRESS, Bonn
Glaudig L (2002) Entsorgungslogistik als unternehmensübergreifendes Konzept. GRIN Verlag
Götz T (2010) SAP-Logistikprozesse mit RFID und Barcode, 2nd edn. SAP PRESS, Bonn
Gulyássy H, Isermann K (2009) Disposition mit SAP. SAP PRESS, Bonn
Hellberg T (2009) Einkauf mit SAP MM, 2nd edn. SAP PRESS, Bonn
Hoppe M, Käber A (2009) Warehouse management mit SAP ERP, 2nd edn. SAP PRESS, Bonn
Iyer RD (2007) Effective SAP SD. SAP PRESS, Bonn
Kirchler M, Manhart D, Unger J (2008) Service mit SAP CRM. SAP PRESS, Bonn
Lauterbach B, Fritz R, Gottlieb J, Mosbrucker B, Dengel T (2009) Transportmanagement mit SAP
  TM. SAP PRESS, Bonn
Liebstückel K (2010) Instandhaltung mit SAP, 2nd edn. SAP PRESS, Bonn
Matyas K (2008) Instandhaltungslogistik. Hanser, München
Melzer-Ridinger R (1995) Materialwirtschaft und Einkauf, vol 2. Oldenbourg, München
Muir N, Kimbell I (2009) Discover SAP, 2nd edn. SAP PRESS, Bonn
Pfohl H-C (2010) Logistiksysteme: Betriebswirtschaftliche Grundlagen, 8th edn. Springer,
  Heidelberg/Dordrecht/London/New York
Scheibler J, Maurer T (2010) Praxishandbuch Vertrieb mit SAP, 3rd edn. SAP PRESS, Bonn
Singh J (2007) Implementing and configuring SAP global trade services. SAP PRESS, Bonn
von Rötzel A (2009) Instandhaltung: Eine betriebliche Herausforderung. VDE-Verlag, Berlin
Wöhe G, Döring U (2008) Einführung in die Allgemeine Betriebswirtschaftslehre, 23rd edn.
  Vahlen, München

Index

J. Kappauf et al., *Logistic Core Operations with SAP*,
DOI 10.1007/978-3-642-18204-4, © Springer-Verlag Berlin Heidelberg 2011